The 21ˢᵗ Century Environmental Revolution

(Second Edition):

A Structural Strategy for Global Warming, Resource Conservation, Toxic Contaminants, and the Environment

/

The Fourth Wave

Book I
Waves of the Future Series

By Mark C. Henderson

Waves of the Future
http://wavesofthefuture.net

Published
by
Waves of the Future

Cataloguing in Publication

Henderson, Mark C
    The 21st century environmental revolution : a structural strategy for
global warming, resource conservation, toxic contaminants, and the
environment : the fourth wave / by Mark C. Henderson. -- 2nd ed.

(Waves of the future series ; book I)
Includes bibliographical references.
ISBN 978-0-9809989-1-7

    1. Environmental sciences. 2. Environmental policy.
3. Environmental economics. 4. Environmental degradation--Prevention.
I. Title. II. Series: Waves of the future series bk. 1

GE105.H45 2010   333.7   C2009-907185-1
Library and Archives Canada

Waves of the FutureSeries
http://wavesofthefuture.net

Table of Contents

Acknowledgment

I would like to express my gratitude to Pierre Bédard who has supported my work and helped make both editions of this book a reality.

Book II
of the
Waves of the Future Series
is due out in September 2010.

Check the publisher's website for release date and best places to buy.

http://wavesofthefuture.net/

Introduction

"Wasters, polluters, and those who do not care for the environment or are unable to adapt would only fall behind and see the fate of the dinosaurs" (Henderson, 2008, p. 83).

This is a quote from the first edition of this book, which was released on June 8, 2008. Less than six weeks later, oil topped $147 a barrel, and the North American automobile industry—happily cruising along on SUV sales—found itself on the brink of disaster. Layoffs and plant closures followed. The SUV was out and the electric vehicle, in. The problem was, the latter was still in development and years away from hitting the production lines.

General Motors and Chrysler would have declared bankruptcy had it not been for bailouts from the U.S. and Canadian governments. Of course, there was the financial crisis, but the two corporations were already in major trouble before that.

Single-handedly, the sharp rise in the cost of energy changed the whole picture for the environment. Global warming was not the only issue of concern anymore. Resource depletion had reached popular awareness. Most economists, and many environmental deniers, now accept the fact that the era of cheap, abundant oil is coming to an end.

Petroleum is highly replaceable as there are plenty of renewable and relatively inexpensive alternatives. It is not the case for metals. We will not weather their shortage easily if at all. The planet's reserves of metallic resources are surprisingly small given the current—and growing—world population. The earth has been around for five billion years. In less than a century of intense consumption, we wiped out the bulk of petroleum reserves! Metals are not far behind. The problem in their case is that in most instances they cannot be replaced with cost-effective renewable alternatives.

In addition to global warming and resource depletion, contaminants continue to be a significant problem. Recent research in the U.S. and Canada shows newborn babies with dozens of toxic compounds and carcinogens in their tissues (Environmental Defence, 2007, December 15; Houlihan, J., Kropp, T., Wiles, R., Gray, S., & Campbell, C., 2005, July 14). Brominated fire retardant chemicals were widely found in mother's milk samples in an American study (Lunder and Sharp, 2003, September 23), and the presence of a rocket fuel ingredient (perchlorate) was detected in nearly half of U.S. states' drinking water. (Rocket fuel, 2004, June 22).

Cap-and-trade will reduce greenhouse gases but does little with respect to the depletion of many resources (metals, fisheries, etc.) and the continued contamination of land, air, and water with several kinds of toxic compounds and carcinogens. We need a more comprehensive strategy, one that addresses not only global warming but also other major issues on the environmental agenda. This is what this book is about.

Of course, the next question is: if we have difficulties dealing with global warming alone, how can we possibly handle other environmental problems at the same time? Firstly, the cost of not addressing issues (contamination, waste disposal, loss of resources, healthcare, etc.) is increasing rapidly.

Secondly, the use of the structural strategy proposed in this book would see problems being dealt with at the source, meaning much greater effectiveness and lower costs.

Thirdly, the approach discussed here is based on a revenue-neutral plan and would essentially be free, i.e. we could have a much more comprehensive and effective strategy than cap-and-trade at a lower cost. If you are looking for the cheapest and most effective deal, this is it!

The question becomes: why are we going for cap-and-trade at all? It is a strategy that is narrow in focus and more expensive, that lets the depletion of resources go on unabated—meaning more crises in the near future—and that allows for the continued contamination of land, air, and water with a range of toxic compounds.

Good question!

The Decline of Modern Civilization

Modern civilization probably reached its peak in the late 1970s and early 1980s, with industrialization at its height and economies benefiting from an abundance of cheap fossil fuels.

Very different factors will define the coming years: disruptions caused by global warming, inflation resulting from the depletion of resources, and higher levels of toxic contaminants and carcinogens in the environment.

We will have to change one way or the other. Global warming is beginning to kill and cost us dearly. It will only get worse as time goes on. Fossil fuels are running out (not to speak of the fact that their combustion produces large amounts of greenhouse gases). As such, shifting to renewable energy is becoming less and less a matter of choice. Metals are being depleted and will follow a pattern similar to that of oil with respect to prices. Either we implement strategies to conserve them now or we will face crises worse than we are seeing with energy.

The current recession provides a break in the trend. Remember the headlines from the summer of 2008 when the price of oil peaked? News reports were talking about a tripling of the price of rice from a combination of speculation, a sudden rise in the cost of energy, and alternative fuels (ethanol) competing with food for agricultural land. Many feared that famine would affect many regions of the world.

Pollution is widespread, hurting us and our children as well as future generations. We are only beginning to pay the price for decades of neglect, be it in the cost of treatment for many diseases (respiratory problems, cancer, etc.), waste disposal, or loss of amenities. World resources such as the oceans fisheries are beginning to be seriously degraded by toxic contaminants.

The fiscal stimulus money recently thrown at the environment is certainly a shot in the arm. However, it is one-time funding and mostly not being spent within the context of a comprehensive and thoughtfully planned long-term initiative.

Cap-and-trade is not enough. It does not address most of the issues on the environmental agenda. Together with the financial crisis money, it only obscures the facts that they both fall sorely short of what is needed and that there are other strategies that could deliver what we need for the environment. Not addressing the problem of resource depletion will mean a chaotic future and economic decline which, in turn, will lower revenues to fund environmental initiatives.

Rather than focusing on ad hoc and trendy strategies addressing only a narrow range of issues, this book looks for more fundamental and structural solutions—ones addressing problems at the source and covering most of the issues on the environmental agenda. With such an approach, we stand a better chance of keeping economies on course while transitioning to a post-modern sustainable society. The comprehensive strategy proposed in this book would make the planet much healthier by the end of the process... all of this coming with the added bonus of revenue neutrality.

Change will happen one way or the other. We can choose to do as little as possible for the environment and pay the costs in terms of erratic, destructive, and deadly weather patterns, skyrocketing resource prices, higher healthcare expenses, and a highly contaminated world. Or, we can opt for a revenue-neutral strategy that would turn a destructive economy into a constructive one and deliver for us and our children a better, healthier, and cleaner world.

The Series, the Book, the Future

For many, Hurricane Katrina and its devastating consequences in 2005 was the first serious environmental wakeup call. Unfortunately, it came after disaster hit.

Today, governments are increasingly moving towards the

curtailment of greenhouse gases to slow down global warming. Efforts, however, are at best halting. As already stated, issues are much bigger than global warming alone. The Kyoto Protocol (of which the last round of negotiation was the climate change conference in Copenhagen, Denmark, in December 2009) is not the most effective way to deal with greenhouse gases and falls short. We need to look at much more profound changes.

A more powerful way to approach today's problems is to make market economies work for the environment instead of against it. The billions of dollars we will eventually spend on problems are really to clean up their mess! It does not have to be like this. With specific changes to the incentive structure of economies, we can make manufacturing greener without major disruptions to society or the market system itself.

Waves of the Future

Decades ago, Alvin Toffler wrote about three waves of change that swept over the world and transformed societies on a massive scale. We are now witnessing the fourth one: a transition to a greener world. Whether it happens grudgingly and destructively, through havoc and reactions to disasters or through forethought and in a progressive manner, it is unavoidable.

Most of us woke up this morning assuming that the sun would rise. Beyond that, it is not so clear what can or cannot be assumed. We know far less than we think about what really lies ahead of us. A decade ago, global warming was considered by many but a theory, the fancy of a scientist's imagination. Now, it is an ugly reality that takes life and destroys indiscriminately, that will starve many, and that we will live with for decades.

Undoubtedly, the 20th century saw a staggering number of technological advances. From a social and economic point of view, many positive developments have also occurred. However, we have failed to address several of the problems that plagued societies during that period. Given enough commitment, many could have been solved a decade or more ago.

We have made some progress in terms of world peace, but we have also developed weapons of mass destruction capable of wiping out life on the entire planet. Not enough headway has been made in the fight against unemployment, social disparities, and world poverty given the vast increase in wealth we saw in the second half of the century. Even the richest countries in the world have failed to eradicate homelessness and dire poverty on their own streets.

In a relatively short period of industrialization, we have succeeded in producing a tremendous amount of pollution, some of which reaching just about everywhere including Antarctica, some of which lasting for extremely long periods of time.

There is a problem with the current economic system, at least with the way it is set up. To create jobs, we need to raise consumption levels. However, this also means increasing pollution and the depletion of non-renewable resources. China and India are viewed as vast consumer markets with a huge potential for job creation. But what of the effects that such massive consumption will have on the environment and resources? China's latest Kyoto commitment is to continue to increase its greenhouse gas emissions. Shockingly, it is heralded in the media as a breakthrough for the accord!

Two things are becoming clear. Firstly, the destructive path we are currently on has to change; our impact on the planet is just too devastating. As such, we cannot rely on narrowly focused solutions: fundamental changes must occur. Secondly, our destructive powers have grown significantly in the last one hundred years. That trend is accelerating as world production, consumption, and population continue to increase.

At the beginning of the 20th century, we did not have the technology to cause a lot of damage. Today, we have our pick of ways to destroy ourselves and the planet: weapons of mass destruction, toxic contaminants, the depletion of essential non-renewable resources, etc.

Looking Ahead

World development has always been determined by both technological and societal advances. Although physical sciences have historically been important, one may think, for example, of the way democracy has fashioned modern societies or the impact of the human rights movement. In the current geopolitical chaos, the world may take a turn for the better or the worse. The stakes are high.

We have reached a point where things cannot continue the way they have in the past. History has shown many an empire crumbling because of their inability to shift thinking and challenge fundamental assumptions when the time had come. In the years ahead, countries will have to decide whether to maintain the status quo and move towards decline and disaster or to challenge their assumptions and trade a little pain now for a much brighter future.

There is generally a lot of resistance to new ideas, and things usually change very slowly. This is often for the better, preventing society from hastily going down disastrous paths. However, we do not have the luxury of time on many environmental issues—unless, that is, we don't care about future generations. This is the time for audacity, not for delays and denials.

Few new and creative ideas were hatched in the last 20 years. For the most part, we are only rehashing left-right political thinking when what is really needed is a fundamental change of approach, one that will restructure the market system and reconcile it with society and the environment.

The Fourth Wave will force us to change. We will act either grudgingly—as slowly as possible and after a lot of destruction has occurred... just as we have been doing so far—or we will reassess the past and define a new agenda for the future, looking at fresh ideas and exploring more fundamental solutions.

The Fourth Wave

The 21st Century Environmental Revolution is the first book of a series which deals with contemporary issues (the *Waves of the Future Series*). Its first part provides a historical perspective on current social and economic issues by going back in time through Toffler's first three waves of change. It also outlines the rationale behind a fourth one which will transform society and the world in which we live either constructively or destructively, depending on the choices we make.

The second part of the book deals with the spectrum of environmental issues and the question of non-renewable resources.

The third and most important part lays out a comprehensive structural strategy capable of addressing most of the issues on the environmental agenda: global warming, non-renewable resources, toxic contaminants, etc. Simplicity, economic feasibility, ease of implementation, and integration to the market system are focal points of the approach.

The strategy articulates together a number of tactical changes to the incentive structure of economies. These would result in a reconciliation of the interests of industry with those of society and the environment. Its use of market forces makes it a tool much more powerful than cap-and-trade and one capable of delivering change quickly and effectively, and of doing so in concert with the economic system.

This book is international in scope. One of the main difficulties in its writing arose from the need to reach out to a broad audience: political changes do not happen without the support of a large number of people. Too many good ideas often die in academic circles. Others take years to break through to popular awareness. We do not have the luxury of time on the environment.

On the one hand, the challenge was to make issues, concepts, and solutions readily understandable to the vast number of those who have political power, the voters. On the other hand, there was a need to also make a solid case to the academic establishment, and this, without getting overly technical and putting off voters.

To that purpose, chapter 2 is designed to be light reading and mostly historical or contextual background information for every-body. Chapters 3, 4, and 5 (a brief review of environmental concepts and issues) are aimed at people not overly familiar with the subject. Academics might want to read chapter 1 and then skip through to chapter 6 where the outlining of the strategy discussed above begins.

The environmental approach proposed in this book could turn the tables on many issues and set us on a new road that would make the massive power of economies work for the environment instead of against it. The other option is to *manage-after-disasters* as we have done so far: wait until resources are destroyed, land is contaminated, people die or starve, or some disaster strikes... and then act.

Nature is finally catching up to us: one way or the other, change will happen. The only choice is how.

Welcome to the Fourth Wave.

1. Non-Technological Revolutions

Entering the Third Millennium

In his 1980 book, *The Third Wave*, Alvin Toffler talks about massive waves of change bringing about a total transformation in the way people work, live, and relate to each other. His work inspired many and became a cult classic in the early 1980s.

In his analysis, Toffler describes waves of change sweeping across the globe at different paces in different areas, leaving some early societies untouched but propelling others into the future. The waves overtook each other, bringing about transformation in their passage. They also clashed together, creating the disruptions associated with the birth pains of new eras.

The tides of change Toffler (1980) talks about began with the First Wave, or the agricultural revolution that took place during the Neolithic period of human history. While the First Wave lasted for thousands of years, the second one—the Industrial Revolution—began around 1750 CE and has already seen its dominance overtaken by the next one. He expected that our generation would fully bear witness to the Third Wave and see it completed within possibly only decades (p. 26).

Although some of his predictions have not come true, many did to a greater extent than he even anticipated. The Internet and the computer age are good examples of this. Even he did not forecast the extent to which they would totally pervade our lives from the inside out and the outside in.

Toffler argued that change was accelerating, a point of view echoed by many others. This phenomenon could have a lot of relevance as the Fourth Wave sweeps across the planet in the years to come.

Scientists are now talking about global warming happening faster than predicted. The sudden spike in the price of oil caught most people by surprise in the summer of 2008. It resulted in very dramatic changes in economies, sharply increasing the price of gasoline, boosting demand for renewable energy, causing the cost of food to rise—dramatically in some cases—and sending the North American automobile industry into a tailspin.

More than any other event in recent history, the July 2008 spike in the price of oil heralds the beginning of a new era of shortages in resources. Although still slow at the moment—and overshadowed by the current recession—the process will accelerate.

Where would we be today—only a couple of years later—had it not been for the slowdown resulting from the financial crisis? The short answer is, in the midst of the summer of 2008 turmoil: expensive fill-ups at the gas pump, the price of goods and food rising from the higher cost of energy and transportation, starvation threatening many parts of the world, etc.

Governments would be under pressure to increase taxes and cut services to make up for higher operating expenses. The worldwide increase in poverty and suffering would mean a greater need for foreign aid and be a breeding ground for discontent, radicalism, and protracted wars—all of which would further strain government budgets. Cap-and-trade would go back to the discussion table, and countries would pull out of the Kyoto Accord or force lower targets.

Environment budgets would likely be the first to face the axe worldwide. Green legislation would be once again rolled back by governments under pressure by right-wing elements arguing that environmental regulations made economies less competitive. Two decades of hard work by environmentalists would go down the drain.

The problem is, even the above will not work this time. While over the last few decades most governments have not had a problem with continuing to poison the environment with toxic compounds and carcinogens, they will not be able to apply the same solution to resource depletion. It would only worsen the problem in this case.

Without conservation and a shift to renewable energy, the price of gasoline will increase faster. The watering down of the Kyoto Accord and environmental regulations will have little effect on this. If anything, it would make things worse by enabling us to wipe reserves out faster.

This is the Fourth Wave!

Technological and Non-Technological Waves

Toffler's first three waves of change arose from technological developments. The fourth one—the transition to a greener world— is very different. It is being forced on us by our own neglect and destructiveness: global warming, resource depletion, contamination, etc.

By the end of the process, we will live in a greener society. What remains to be seen is what condition the world will be in by then: highly damaged (many resources depleted, others destroyed as a result of contamination, high levels of toxic chemicals and carcinogens in our homes, cities, and the environment, much lower standards of living, international tensions, terrorism, etc.) or in a much better shape, with more resources left, lower levels of contaminants, and thriving green economies.

The single-focus (cap-and-trade) and the manage-after-disaster approaches lead to the first scenario. The strategy developed in this book offers a different avenue, one leading to a much greener and cleaner world and one that might be seen a few decades from now as ushering in the 21st century environmental revolution.

The Right Recipe for a Revolution

Major changes or revolutions do happen given certain conditions. Technology was a major factor in bringing about Toffler's Third Wave, the Information Age. It will play a role in addressing today's environmental problems, but the road to a greener world lies at the organizational level, which is the source of the problem. We need to cure the disease rather than just treat its symptoms as we are doing now.

Revolutions happen as a result of a combination of elements. One of the key ingredients is obviously new ideas or solutions. It is what this book is about.

Another factor is popular support. Current levels of political commitment to environmental protection can take us some distance but are often dependent on disaster striking and people falling ill or dying. Short of that, little happens. In addition, popular support varies greatly depending on economic cycles, gains being made in good times and losses incurred in bad ones.

In this new era of resource depletion, most economies will likely decline unless we find alternatives to the single-focus and manage-after-disaster approaches. We will face a long-term process of shrinking resources, rising consumer prices, and decreasing standards of living. This will serve to erode funding and political commitment for the environment. As such, a strategy that does not address the problem of resource depletion is bound to fail, and popular support alone cannot win the day for the planet.

Timing is very important. For example, winning the right to freedom of speech made many social changes possible. Before it happened, movements such as the sexual revolution, gender equity, and unionizing were mere possibilities. After it did, they became virtual certainties. What were mere ideas at first became powerful and unstoppable forces for change. With global warming pressures continuing to mount, the choice to act is becoming less and less an option. The time is now. This is our chance to do it right.

Fundamental legitimacy is also a key element to create a successful revolution. Many social movements overcame incredible odds primarily because of their rightfulness. The fundamental legitimacy of addressing environmental problems is unquestionable, especially now that we are starting to pay the price for our neglect.

Given the right ideas, timing, political support, and fundamental legitimacy, powerful movements can come to life and change the world. The time is ripe for an environmental revolution, but it will not happen via cap-and-trade and manage-after-disaster strategies. We have to look at more profound and fundamental solutions.

A structural approach is what we need.

The Master Plan for the New Millennium

What would we like to achieve in the next century? Do you recall hearing of any plan for the long-term future? The extent of planning for this millennium is often limited to coping with emergencies as they come up—not before they do.

Of course, the Kyoto Accord is somewhat of an exception, but before we celebrate our achievements, we need to consider the facts that it is still fairly shaky after over a decade of existence (with countries moving in and out and China signing up on the condition that it is allowed to continue increasing its emissions!) and that it leaves the majority of issues on the environmental agenda unaddressed. The Kyoto Accord is a band-aid, not a plan for the future.

Our Legacy

What kind of planet will our children inherit from us? Will it be better or worse than the one that was bequeathed to us? What will be our final legacy?

There has been a certain amount of legislation made with respect to the environment but not nearly enough, judging from the lack of progress on global warming, the absence of a plan to address the problem of resource depletion, the pervasiveness of contaminants in the environment, the continued wastefulness in the packaging industry, the intensive use of chemicals in the agricultural industry, etc.

Fresh water is considered a health hazard in many places to the point where it is even dangerous to swim in, let alone drinking it! No matter where you go, many pollutants are found in increasing levels in many fish species.

The issue of mercury, for example, has been around for several decades. Yet, the Mercury Policy Project (a U.S. initiative to reduce exposure to the metal and eliminate its uses) reported that an American study which had found mercury "in the blood of 2% of women in 1999... [detected it] in 30% of the women by 2006" (Mercury Policy Project, 2009, September 1). A separate analysis of fish and water streams across the U.S. from 1998 to 2005 found "27% [of sampled fish] exceeding levels safe for human consumption" (Mercury Policy Project, 2009, September 1).

Increasingly, studies find that our own body tissues and those of infants and children are laced with dozens when not hundreds of toxic compounds and carcinogens. The milk we feed our kids, the water they drink, the food they eat are full of contaminants. Many of them will cumulate over time and progressively poison the environment in which we live.

The news headlines and media focus on climate change overshadow the fact that the contamination of land, water, and air continues unabated. While much attention is paid to sudden and dramatic deaths caused by global warming, little is made of the slow and chronic poisoning of the planet and everything and everyone that lives on it.

Non-Renewable Resources

Few speak of the depletion of non-renewable metallic resources. Everybody has heard about saving trees and global warming, but how much of the earth's mineral resources are we going to leave future generations? The metals upon which much of our lives and lifestyles depend are not renewable. Generally speaking, once gone, they are gone. They do not grow back.

We will start having shortages within a decade or two! The problem is actually acutely critical. Yet, the only planning with respect to non-renewable resources is a conspiracy of silence: take as much as you can, and don't ask any questions or raise the issue. Metals and other non-renewable resources are world capital that belongs to other generations too. We will soon begin paying the price for our own neglect... just as we have already started doing on account of oil. Nature is catching up to us!

The Fourth Wave

We can continue to react to crises and then deal with the aftermath of disasters and increasing environmental destruction, or we can restructure economies to make them green and work for us.

Global warming is only one of the many issues that have to be addressed. The Fourth Wave will force a transition to a greener society. Just how much resource depletion, contamination, and destruction will have occurred by the time this happens depends on us.

We are not doing so well as stewards of the world we inherited. We'd better get our act together soon, otherwise, there won't be much of a planet left for our children.

2. Toffler's Waves of Change

The Pre-First Wave

How would you define our society? Perhaps, looking back at where we come from will help in answering this question. As this book is partly rooted in Alvin Toffler's works, his own descriptions and definitions will be used.

Pre-first-wave societies go back over 10,000 years. Toffler describes them as small bands of nomads living off the land by fishing, hunting, and gathering wild plants. They still exist today in some parts of the world—for example, some tribes in the tropical forests of the Amazon River and Papua New Guinea—but, generally, their way of life gradually came to an end when the First Wave —the agricultural revolution—began taking hold about 10,000 years ago (p. 29).

Pre-Agricultural Life

The pre-agricultural era is generally defined by anthropologists as the first part of the Stone Age—the Old Stone Age or the Paleolithic (500,000 BCE to about 10,000 BCE). Early Stone Age people were nomads who lived off the land. They constantly moved in pursuit of food. They followed the herds of animals they hunted and moved away from an area once it was depleted of game or the plants (fruit, berries, roots, leafy greens, nuts, grains, etc.) they ate (Beers, 1986, p. 21). Their diet changed depending on food availability and season of the year.

Their social structure was generally simple. Their way of

production was not very effective. Because they migrated regularly, everything they owned had to be carried with them. As a result, they could not accumulate food surpluses and only produced what was needed. When hard times hit—as a result of droughts or other natural causes—they simply starved.

The First Wave: The Agricultural Revolution

Toffler describes the First Wave as having begun with and been driven by the agricultural revolution. He locates it in time between 8000 BCE and 1650-1750 CE. Farming and land cultivation began in the first part of the New Stone Age (Neolithic), which ended around 3500 BCE.

During that period, people gradually moved from hunting and gathering to herding and agriculture. That span of time saw the domestication of wild animals, for example, dogs, sheep, and goats. Vegetables and grains were grown and harvested. Different regions of the world saw different types of food being cultivated, depending on the suitability of soils and climates. Rice and yams were grown in Asia. Wheat, oats, and barley were staples of the Middle East and Africa. South America's traditional crops were maize and beans (Beers, 1986, p. 22).

The transformations that the agricultural revolution brought to the Stone Age way of life and the implications it was to have for the future of society were staggering. The adoption of agricultural practices led to greater and more secure supplies of meat, grains, and other foods. The risks of starving from having a bad hunting season were lessened, and people no longer had to move constantly. Reserves existed in the form of herd stocks and grain stores.

Cultivation required the development of a number of new tools and the building of facilities for the storage of seeds. It necessitated land, which not only had to be fertile but also needed to be made ready or suitable for food growing; it had to be cleared of trees and root systems, leveled and drained, tilled, etc. That led to the abandonment of the nomadic way of life. As facilities and cleared land could not be moved around and involved a lot of work, people began settling down around them.

The Passing on of Wealth

A significant economic aspect of agriculture was that wealth or capital could be built over time and passed on to the next generation. Stores of grain, land, equipment, a herd, and facilities could be accumulated over the years and bequeathed to the following generation.

This meant that not only food but also wealth and power could accumulate over the generations. As a result, certain families, clans, and tribes emerged as wealthier and more powerful. The economic surplus and the ability to pass it on laid the groundwork for a social hierarchy to emerge.

Toffler (1980) talks of a generally parallel evolution in all major civilizations—be it in Europe, Asia, or Latin America. Land had become the basis of society. Life—economic, cultural, familial, and political—revolved around it. The village became the center of the social organization. They all saw the emergence of basic work specialization (division of labor). A rigid and authoritarian system of social classes (the nobility, priesthood, military, peasantry, etc.) began to appear. Rank was determined by birth, not merit (pp. 37-38). Neolithic economies were decentralized, and communities were for the most part self-sufficient.

The Rise of the City

The other important development that happened as a result of increased farming and agricultural surpluses was the rise of the city. Large settlements first appeared around 6000 BCE, but the actual urban revolution that marked the beginning of civilization occurred later, circa 3500 BCE. The Tigris and Euphrates rivers in the Middle East, the Nile in North Africa, and the Indus and Yellow rivers in South Asia and Asia were the cradles that gave birth to civilization.

Cities were largely oversized agricultural settlements at first. With the growth of food surpluses, they were able to support a larger non-agricultural class. As the need for tools and technology increased, a new urban class grew, the artisans. They were the forerunners of modern tradespeople.

While today the percentage of people involved in agricultural activities in North America is only in the order of 2% to 3%, virtually all members of a tribe were involved in food generation.

Then, the Second Wave hit.

The Second Wave: The Industrial Revolution

For Toffler (1980) the Second Wave is defined and was brought about by industrialization. It began around the middle of the 18th century, but some of its precursors go back further: oil drilling on a Greek island as early as 400 BCE, the existence of money and exchange, the network of trade routes from Europe to Asia, the emergence of urban metropolises in Asia and South America, etc. (p. 38).

Prior to the industrial revolution, most people were essentially self-sufficient; they produced what they consumed. With the Second Wave, peasants lost their autonomy, and production became intended for markets. Toffler (1980) argues that this created "a way of life filled with economic tensions, social conflict, and psychological malaise" (p. 53).

The Second Wave gave rise to the division of labor, which allowed workers and trades people to be better trained, hone their skills in apprenticeships, and gain more experience in what they did. The division of labor culminated with the invention of the assembly line, a production technique which consisted in splitting up an elaborate process into simple repetitive tasks that could easily be performed individually by unskilled labor. Henry Ford is the name most often associated with its introduction in America. The new approach greatly accelerated production and reduced costs.

The greater affordability of goods translated into increased demand, which paved the way for another Second Wave phenomenon and defining characteristic of industrialization: mass production or the large-scale manufacturing of items that are identical. This drove prices further down and gave rise to today's style of consumerism.

Industrial Technology: The Power to Move Mountains

One of the key elements in the development of industrialization was the multiplication of human strength and power by several folds.

For example, to exploit the tar sands in Alberta, Canada, we have created monster trucks having carrying capacities of 360 tons. In comparison, the average vehicle used to deliver landscaping soil to your home carries loads of 1/2 to 3 tons.

The piece of technology responsible for our ability to move mountains is the engine. Just like the water wheel and the windmill early on, the invention served to multiply labor output. It powered a new era, revolutionized transportation, and eventually created a booming automobile industry. The invention of the engine resulted in a massive increase in wealth and productivity.

However, unlike earlier technology that used clean and renewable energies such as the wind and water, the new contraption was fed by fossil fuels produced in ancient times. This signalled a significant shift away from clean sources of power and represented perhaps our first step down the road towards global warming.

The fossil fuels that we have been tapping into to support our lifestyles for the last century have been a boon but also a scourge in their massive contribution to pollution and global warming. By the early 1970s, the petroleum industry had been so successful in expanding wealth that virtually the entire world economy had become highly dependent upon it.

This is when the first oil crisis hit.

The Third Wave: The Digital Age

Toffler (1980) argues that up until 1973 industrialization had ruled unchallenged. The Soviet Union and the U.S. were locked in the Cold War, both vying for allies worldwide, seeking to expand their influence and shore up their defenses. Multinational corporations emerged as a third power—often under the protection of their national governments—spreading their tentacles across the planet in their relentless drive for cheap resources and greater profits.

Drunk on cheap and bountiful oil supplies from the Middle East, the developed world saw growing stability and unlimited economic expansion. The new wealth set an upbeat mood. Until 1973, that is, when everything came to a screeching halt.

OPEC

August 8, 1960, argues Toffler, might be symbolic of the final stages of the Second Wave. On that day, in a bid to increase profits, Monroe Rathbone, Chief Executive Officer (CEO) of Exxon Corporation, made the decision to reduce royalties on imported petroleum. The other major players in the industry followed suit within days. Oil producing states—many of them, developing countries—were hit especially hard by the losses. They organized an emergency meeting in Baghdad to address the issue. On September 9, 1960, the Organization of Petroleum Exporting Countries (OPEC) was born.

For many years, it had little success in raising oil prices. However, taking advantage of the outbreak of the Yom Kippur War in 1973, OPEC was successful in cutting production and increasing revenues several folds.

Second Wave production systems were highly concentrated and non-diversified with respect to energy. Transportation was almost exclusively petroleum based—and still is very much so today. The industrial base of most countries also generally revolved around the same fuel. The price of gasoline spiked up, sending world economies into inflationary spirals.

In 1980, just as countries were emerging from the first oil crisis, OPEC struck again. The cost of a barrel of oil tripled suddenly to above US$ 35. The rise in prices resulted in austerity measures being implemented in both developed and developing parts of the world. The new economic conditions brought some countries to the very brink of bankruptcy. In the years that followed, world oil prices dropped back to lower levels.

Other cartels do exist. However, none had struck so close to the heart of the industrial economy.

Today, the price of petroleum is up to new heights. It peaked past US$ 147 a barrel in July 2008 and has come down since, but it is expected to climb back up as the world economy edges its way out of the current slowdown. Because of the high inflation levels of the last two decades of the 20th century—not to speak of the drop in value of the U.S. dollar—today's prices are not as high as they seem to be. For example, in 2008 when the cost per barrel reached over $120, the news media reported that level to be comparable to what it was after the second oil crisis. Of course, that is little comfort to drivers and consumers.

Toffler's De-Massification

Second Wave industry developed from mechanical inventions and focused on mass production. With the advent of the Third Wave, the Digital Age, Toffler saw the pendulum swinging back from the petroleum-based heavy industry towards more appropriate-scale and less oil-intensive technologies. Computers and the Internet would lead to a de-massification of industry: decentralized operations, increased customization, smaller production runs, etc.

He also makes the case that the Third Wave industrial base would operate on a more sustainable basis and be composed of a mix of *high-stream* industries (large scale and science based) that would be under tighter social controls and friendlier to the environment, and of *low-stream* industries—of smaller, more appropriate, and human scale—that would rely on new and sophisticated technology.

The Missing Link

Despite Toffler's valiant efforts, we do not have yet the answers to all the questions posed by the 20th century. One of the most important issues in socio-economic change is implementability. No matter how good something looks on paper, it has no use if it is too expensive or impractical. Neither has it much value if it is not powerful enough to solve a given problem. An insufficient dose of medicine will generally fail to cure a disease and can actually be harmful and result in the patient's death.

It should be obvious to all of us that with respect to the environment there is still a piece of the puzzle missing. Despite the scientific advances of the last century, we are not keeping up with problems, let alone solving them.

The earth and the atmosphere are slowly being poisoned by a number of pollutants. When people do not get sick or die, the contamination continues silently, unhindered. Environmentalists' efforts are laudable, but they are not enough. Problems are massive and will not get resolved with piecemeal or incremental solutions. This is where this book comes into play. It brings in new thinking and takes an appropriate-scale and comprehensive approach to problem-solving for the environment.

The New Millennium

At the beginning of this new millennium, oil remains an overriding concern for modern societies. If we continue down the same path and do not address the problem of depletion, the future will be one where scarcity will turn one non-renewable resource after another into objects of power. This will dramatically raise international tensions and fuel endless conflicts. Poverty will progressively increase in both developing and developed countries.

Unless we change the way we manage resources, we will be looking at a very chaotic future, and the current turmoil will only be a taste of what is to come.

3. Energy: The Past and the Future

This chapter provides a general introduction to energy issues and will serve as a basis for the development of the environmental strategy proposed later on. If you already have an advanced knowledge of the topic, feel free to fast forward through this part or jump to the next chapter.

Ancient Energy

Two defining characteristics at the core of contemporary society are technology and fossil energy. The massive productivity of modern machinery is not only the result of engineering designs but also of energy. The work that machinery produces does not come from human muscles but from fuel.

A lot of today's industrial technology is based on petroleum. One of its distinctive characteristics is that it is a fossil fuel. Oil, like other mineral resources, comes free from the earth. What we pay for are the costs of extraction, refining, and distribution (getting it to the gas pump).

We are vastly more successful than Stone Age people not only because of the technology we have developed but also from tapping into vast and inexpensive sources of energy produced in ancient times. When the price of oil increases as has been happening, we see how quickly it can affect our standard of living and how much of modern society's success depends on cheap fossil energy: the cost of everything goes up, some countries face food crises, and a lot of wealth simply disappears.

The Carbon Cycle

Oil, coal, and other fossil fuels come from biological (plant and animal) matter or *biomass*. Over millennia, vast amounts of vegetation and animal wastes were deposited at the bottom of bodies of water or submerged as a result of one cataclysm or another. Under certain conditions, part of the sediments eventually turned into coal, petroleum, natural gas, or other fossil fuels. In areas of the world where non-permeable layers formed on top of the biomass, even the more volatile elements, such as natural gas, remained trapped.

The question is, what is the vegetation that turned into oil made of? Most of us would say dirt, but we would be wrong. If you remember your biology lessons, you know that the bulk of the weight of a tree is composed of only a small fraction of soil. The two largest components of plant matter are water and carbon, as in carbohydrates or what the body uses to produce energy.

The latter does not come from the soil but from carbon dioxide gas (CO_2), which is a natural component of the atmosphere. It is one molecule of carbon attached to two of oxygen. Breaking these apart takes energy. Recombining them gives energy. With power (light) from the sun, leaf cells break apart the molecules of the gas and use the carbon for their own growth in a process called *photosynthesis*. That element is the main building material of plants. The oxygen is released into the air and left behind for us to breathe.

When vegetation decays under specific circumstances, it is transformed into hydrocarbons—the main components of petroleum. It is another way in which the carbon originally captured from the air by plants is stored in organic matter.

When you eat and burn the carbohydrates from food in your own body, you recombine carbon and oxygen to reform carbon dioxide, which you breathe out. The process releases energy that fuels the muscles and other functions in your body. In the same way, hydrocarbons are burnt in car engines in a process that recombines carbon to oxygen from the air (combustion). The carbon dioxide gas produced is released back into the atmosphere via car exhausts. Of course, as petroleum is not pure hydrocarbon, many pollutants are also created and vented out through the combustion process.

Global Warming and the Carbon Cycle

Carbon dioxide is a greenhouse gas, or a gas that significantly contributes to global warming problems. It is also one of the main targets for reduction under the Kyoto Accord. As we should all know by now, it acts as a blanket around the earth and reduces heat radiation into space, keeping the planet warmer. Higher concentrations of greenhouse gases in the atmosphere result in rising global temperatures that threaten to melt the polar ice caps, raise sea levels, flood coastal regions, disrupt global weather patterns, and even trigger a new ice age.

Vegetable oils are also a form of fuel. They burn like petroleum and can be processed to produce biodiesel which can be used in regular engines. Whether you burn petroleum, coal for electricity generation, wood, or biodiesel, you always end up recombining carbon with oxygen from the air to re-form carbon dioxide, which is emitted back into the atmosphere. Growing vegetation reduces global warming. Breathing and burning fuels, on the other hand, increase it.

The carbon dioxide that humans and animals breathe out is not what creates global warming problems. Neither is it the burning of wood. What we eat and the logs we burn are carbon that has recently been taken from the atmosphere by vegetables or trees in their growth process. We are just putting it back in. These activities are *carbon neutral* because of that: one ton of CO_2 removed plus one ton added equals zero. Based on the same principle, renewable fuels generated from corn or other crops are technically carbon neutral. In practice, their production does currently involve large amounts of gasoline or diesel, making the process far from carbon neutral.

Fossil fuels are the major culprits behind the greenhouse effect. Today's large-scale mining of coal and extraction of oil release massive amounts of carbon from deposits that were produced millions of years ago. This results in increasing concentrations of carbon dioxide in the atmosphere and rising global temperatures due to its blanketing effect. The problem is worsened by the fact that we are emptying vast pools of ancient energy in a relatively very short period of time: decades, perhaps a century or two.

Another significant source of greenhouse gases is the large-scale and permanent deforestation of the planet. As seen earlier, trees store carbon. Growing one and cutting it down for fuel is carbon neutral: plus one minus one equals zero.

However, the total forestation on the planet is itself a vast reserve of carbon just like the underground pools of oil. Although a single tree may live or die, the forests themselves have been around for millions of years. Reducing global forestation without replanting adds new carbon to the atmosphere and contributes to global warming. Worse, decreasing the total amount of forested area around the globe has a secondary effect: it also reduces the planet's ability to take carbon out of the atmosphere.

The world's oceans play a significant role in carbon absorption. Analyses show that oceanic waters have been able to absorb about half of the carbon dioxide emitted as a result of human actions (*anthropogenic global warming*) since the beginning of the Industrial Revolution. This has served to slow down climate change.

However, the absorption has had a dramatic impact on marine life. As carbon dioxide is taken up, it transforms itself into carbonic acid which increases the acidity of water and removes calcium carbonate from oceans. The latter is needed for the constitution of shells of many marine species, some of which form the very basis of the oceanic ecosystem. Plankton, for instance, is at the bottom of the food chain and is critical to the survival of many species. Its reduction can have devastating effects throughout the marine ecosystem. This also implies that global warming problems could not be solved through the injection of carbon dioxide into oceans (Lean, 2004, August 1).

Transition Fuels

Because fossil fuels are used so massively in today's society, most of the experts in the field do not believe that a full direct conversion to renewable energy is going to be possible. They talk about transition fuels that are less carbon intensive and that could be used in the short and medium term. Here is a brief look at them.

Coal, which is abundant in many countries, is being investigated as an alternative source of energy for the future. It is already widely used in the industry for steam, heating, and the production of electricity. As fossil fuel, it is not renewable or carbon neutral. Its combustion is a source of mercury and other types of pollution around the world and contributes to global warming.

Cleaner coal technology is being researched. The production of liquid fuels (for example, diesel through the Fischer-Tropsch process) and extraction of hydrogen from coal are possible future avenues for the exploitation of this resource.

The industry is also looking at carbon capture and sequestration (CCS) as a means to making this fossil fuel a viable alternative for the future. This avenue generally involves pumping emissions underground. At this point, it is risky and unproven, with the potential for leakage, ground water contamination, and creating geological instability.

Contrary to news headlines, *clean coal* does not exist and may never do so. Its potential for being a transition fuel will greatly depend on how clean it gets with respect to not only greenhouse gas emissions but also its mining and extraction as well as other pollutants related to its use.

Nuclear energy has seen a renewal in the wake of the oil crises. However, like metals, fissile materials are minerals that are depletable. Furthermore, we still do not have any fully safe options for long-term storage of the radioactive waste produced by the industry.

Spent fuel poses a security threat as it can be used to build dirty bombs (standard explosives packed with radioactive material). Although these would not set off a nuclear explosion, they could contaminate a wide area. Multiplying the number of reactors worldwide would also increase the risks of meltdowns. Nuclear energy is cleaner but not a clean option per se.

The use of natural gas has been increasing over the last decades and is expected to continue to do so. Growing interest in this source of energy and more exploration have resulted in an increase of known reserves. Although natural gas is not a renewable energy and is depletable, it is many experts' best hope as a bridging fuel. It has good prospects for enabling us to make the transition from petroleum to the renewable energy sources that will power our future.

Natural gas burns much more cleanly than gasoline, diesel, and heating oil. Its combustion also produces less greenhouse gas than other fossil fuels and does not release any sulfur dioxide—a toxic agent—or particles. Furthermore, the emissions of nitrogen oxide in gas-fired power plants are much lower than those of newer coal technology for electricity generation (Geller, 2003, p.25). Those of carbon dioxide in natural gas power plants are less than half (55% to 65% lower) of those of their coal power equivalents (Geller, 2003, p. 25).

Natural gas is more widely distributed on the planet than petroleum, thus decreasing the world's dependency on Middle East oil. A number of mega-projects for the development of new sources of natural gas are under way in many countries around the world. The sheer size of the capital investments already involved may simply preclude turning back the clock on natural gas even though it does contribute to global warming. Next, we will take a closer look at renewable energy options.

Renewable Energies

Renewable energy is a vast and quickly evolving field of research. It could be in itself the subject of an entire book. This section provides a brief overview of the variety of renewable energies currently in existence and of some of the issues relating to specific resources.

Hydroelectricity, a relatively clean and renewable energy, is generally produced by damming rivers and is more abundant in some countries than others. However, it is not a new form of energy. Hydroelectricity production saw an expansion as a result of the world oil crises of the 1970s and 1980s.

It is expected to continue to grow in the future. However, its expansion is limited by the availability of sites and the impact of hydroelectric projects and their construction on the environment (damage to fish habitats and spawning routes, release of mercury, etc.).

An important aspect of this renewable energy is that dams have a limited lifespan. Reservoirs eventually fill up with silt, rendering projects uneconomic after 60 to 100 years on average.

The future of hydroelectricity may lie in smaller scale technologies —which are more environmentally sound and less disruptive to sport and commercial fishing—as well as in measures to prevent siltation.

Biomass is also an old form of renewable energy. It accounts for a significant amount of primary fuel and power production in many countries and comes in different forms, of which some are old and others, new.

Wood has been used directly as combustible since the domestication of fire. To this day, it is still used for heating as well as electricity production. The technology has evolved, but the principle is still essentially the same: combustion. Slow-burning stoves, wood pellets made from byproducts of the lumber industry, and external air intakes combined with heat exchangers are some of the new technologies designed to make combustion more efficient and take advantage of plentiful and cheap leftovers from the forestry and other industries.

Increasingly, wood derivatives and residues are also used for low-grade heat, steam, and even for the generation of power in the lumber industry itself. Wood is plentiful in many countries. It is a renewable resource as long as the industry is properly managed. It is carbon neutral but not really a clean energy.

Having a few houses in the countryside use wood stoves is one thing, but an entire city doing so is something else altogether. It would vastly add to the smog and pollution problems that already exist in many urban areas. Montreal, Canada, sees many days of *winter haze* on account of the increasing use of wood stoves. Some small towns in Ontario have been experiencing air pollution levels similar to Toronto's for the same reason.

New advanced catalytic combustion stoves are believed to reduce particle emissions, smoke, and other pollutants by about 80% compared to earlier models. Whether or not this new technology will make feasible their use on a large scale remains to be seen.

Also included in the biomass category are solid wastes from both municipal (garbage) and industrial sources. These can also be burnt to generate steam, heat, and electricity.

Some European countries are currently making plans for building several wood-burning power plants as part of their Kyoto

Accord commitment to reduce greenhouse gases. The operation would involve importing wood—which is technically carbon neutral—from Brazil and other countries.

It is unclear, however, how much better that would be for the environment. The long-range transportation involved would not be carbon neutral, and the scheme would amount to trading non-toxic greenhouse gases like carbon dioxide for toxic pollution (many of the byproducts of the combustion process). The approach would also increase pressure for deforestation and the replacement of biodiverse tropical forests with non-biodiverse tree plantations. This is an instance of what can occur with single-focus strategies: a problem is fixed by creating another, or *problem displacement*.

Biomass can also be used to generate bio-gas. Methane—an alternative to natural gas—is increasingly produced from a variety of organic byproducts and leftovers from industry and other sources. The gas can be collected from landfill sites, generated from municipal sewage, or produced through anaerobic fermentation of agricultural crops and byproducts.

Gasohol, ethanol blended diesel, and E85 are other forms of biomass energy. Gasohol is a mix of about 90% gasoline and 10% ethanol, or regular drinking alcohol. It is cleaner and deemed to be a superior fuel for winter driving. Ethanol can be produced from several types of crops: for example, wheat, barley, and corn. Unfortunately, these are the same grains that feed people around the world. As we saw in the summer of 2008, the production of biofuels using prime agricultural land and crops can have a dramatic effect on the price of food.

Ethanol blended diesel contains 10% ethanol. It is cleaner burning than the fossil fuel alone. Both gasohol and ethanol blended diesel can be used directly in regular engines.

E85 is 85% ethanol with only 15% gasoline. Its use requires engine modifications. Some U.S. automakers have already begun to make new motors that can burn either gasoline, gasohol, or E85.
The production of liquid fuels from agricultural residues and non-food crops grown specifically for energy (for example, switchgrass and trees such as willow and poplar) has attracted growing interest. Some of the issues related to biofuels are the initial capital costs

(equipment, facilities, etc.), the development of a common distribution system (i.e. gas stations), and the competition with food crops for land.

Biodiesel is an alternative to regular diesel. It can be made from vegetable oil. Many crops are suitable for its production: soy, sunflower, canola, hemp, etc. It can be used in pure form or blended in different proportions with regular diesel. This alternative can often be fed directly into existing diesel engines with little or no modifications. Unlike fossil fuels, the vegetable oil alternative is carbon neutral and does not contribute to global warming. It also burns more cleanly. Biodiesel combustion produces fewer gas and particulate emissions and lower levels of carcinogens (Pinderhughes, 2004, p. 176).

Wind provides one of the fastest growing and most promising sources of clean and renewable energy worldwide. Turbines are erected in fields to capture the energy from the wind and transform it into electricity, often feeding it directly into the existing electrical grid. However, this type of energy is only available on an intermittent basis. Without air movement or wind, no electricity is produced. This somewhat affects its usability in that power has to be stored or supplemented from another source. Despite this limitation, several countries already have a number of wind farms and are planning for many more.

There are also a number of smaller players in the renewable energy field. For example, tidal power—energy captured from waves and the rise and fall of tides—is clean and renewable but is obviously limited to coastal areas. Geothermal power, or energy extracted from the earth's crust, has fair prospects in a number of locations around the world. It is essentially the same energy as that responsible for hot springs.

Earth-energy systems (EES) are a slightly different approach for the exploitation of geothermal energy. Heat pumps extract low-grade energy from large earth subsurfaces that are warmed by the sun or underground water sources. Conversely, if cooling is needed, heat is extracted from a room and sunk below the ground. EES systems are capital-intensive technologies.

Solar power is probably the most well-known source of renewable energy. It attracted attention when people first began considering alternatives to fossil fuels following the oil price hikes of the 1970s and 1980s. It is currently used for both space and water heating in domestic and industrial sectors worldwide.

There is a variety of systems for capturing energy from the sun. They fall into two categories: passive and active. Passive solar systems primarily capture heat from the sun through windows or sun-absorbing matter for direct space or water heating. Historically, windows were not energy efficient. They provided lower insulation than walls. However, new technology that offers better heat-loss prevention and lower emissivity has improved windows to the point where advanced models now provide a positive energy supply.

Solar walls and green roofs are other instances of passive technology. The former consist, for example, of perforated metal paneling designed to absorb solar energy from south-facing walls on buildings. The technology is usually used for air and hot water heating as well as cooling in industrial plants. It is gaining grounds in renewable energy markets because of its cost-effectiveness and relative ease of installation. It can provide up to one third of a building's heating and air make-up needs. Solar walls are also being investigated internationally for use in other applications such as commercial and agricultural drying.

Green roofs are usually flat rooftops covered with live greenery. They provide passive insulation against heat in the summer and cold in the winter. Other benefits include the reduction of stormwater runoff and lowering of urban temperatures in the summer. They are relatively low tech and are quickly gaining in popularity.

Solar photovoltaic energy (PV) is an active type of technology. Semi-conductors are used to transform sunlight directly into electricity. PV has historically been too pricey to replace conventional sources of energy, but its economics are changing. It has uses in distant areas where there is no existing electrical network and in stand-alone applications. Examples of these are residential locations where grid extension would be expensive, road signage, coastguard systems, and remote monitoring. The main challenge

for PV as an energy for the future is cost reduction. New technological developments and production automation are avenues
through which more competitive prices may be achieved.

Concentrating solar power (CSP) indirectly produces electricity
by focusing sunlight with parabolic mirrors or fields of mirrors onto
a small surface to produce steam to power turbines. An example of
this is the PS10 solar power tower in Seville, Spain, in which a
semi-circular array of mirrors directs sunlight to the top of a multi-
story structure where steam is generated. Both PV and CSP are
increasingly the focus of larger scale operations.

There are many other solar technologies—both passive and active
—in use and in development for various applications: electricity
generation, residential and commercial water heating, and air make-
up.

A Concrete Example of the Renewable Energy Potential

Brazil has millions of cars running on ethanol or one of its blends.
This came as a result of a government program aimed at decreasing
the country's dependency on foreign oil and producing energy
domestically from sugar cane, a local and plentiful crop. The
National Alcohol Program (Pro-Alcool), as it is called, was insti-
tuted following the first oil crisis. Thanks to the initiative, Brazil
has been a net energy producer since 2006 (Gerson Lehrman
Group, 2009, July 16).

Part of today's U.S. transportation could be powered by energy
produced by Americans to the benefit of the entire country. Brazil
—a country much less well off—has been doing it for decades,
proving that renewable energy strategies are feasible. Of course,
the use of prime agricultural land for the production of biofuels is
becoming less and less of an option, but there are alternatives.
Ethanol, biodiesel, or methane (the main component of natural gas)
can be produced from agricultural surpluses, crop residues,
garbage, industrial waste, sewage, manure, and even cellulose
(wood).

From an economic point of view, the National Alcohol Program
enabled Brazil to plow back its wealth into its own farmers' fields,
creating jobs and boosting its local economy. The strategy paid off
handsomely. During the same period (since the first oil crisis), the

U.S. took massive amounts of American wealth and plowed it into the Middle East, enriching countries like Saudi Arabia and impoverishing itself by trillions of dollars.

Many renewable energy technologies have existed for a long time. In many cases, they already made a lot of economic sense decades ago. Had the world followed the example of Brazil when it went through its own transition following the first oil crisis, global warming might not be the issue it is today, and most countries—including the U.S.—would be closer to self-sufficiency in energy, and likely much better off economically.

Renewable Energy Issues

Renewable energies have significant advantages over fossil fuels in that they are generally cleaner, unlimited, and do not contribute to global warming problems. However, many are not as concentrated or portable as petroleum. These two qualities are important factors in their potential for replacing fossil fuels in the transportation industry. Electrical power may be suitable for commuting to work, but storage capacity is still low at this point in time, making it impractical for heavy-duty and long-range transportation. There may also be temperature issues in cold climates.

Technological development is not the most important obstacle to the widespread and large-scale use of alternative energies, to their replacing fossil fuels as power base for economies around the world. Up until very recently, ups and downs in the price of oil were the problem. We saw bursts of R&D (Research and Development), government subsidies, and investment in alternative fuel technologies in the wake of the oil crises of the 1970s and early 1980s. Many of these were subsequently lost when the cost of petroleum dropped.

Several types of renewable energy are only profitable when the price of oil remains high. As such, unpredictable petroleum prices were a stumbling block for the industry in the last three decades. This is now changing as most experts in the field believe that the cost of oil will remain fairly high and trend upward. A more recent problem is the rise in the price of food from the use of edible crops and good agricultural land for the production of biofuels. There are ways to address the issue, but as the world's population continues to grow, the pressure on land and crops will likely remain an ongoing concern.

Another issue is that converting to new energies often requires a substantial investment in infrastructure. For example, gas stations need to be modified to accommodate new fuels or a new distribution infrastructure might have to be built parallel to the existing one.

Methane Hydrates: Energy of the Future?

Methane hydrates have been heralded by some as a potentially major source of energy for the future. They are gas molecules trapped in ice on the ocean's floor. Research is still very preliminary at this point in time. Reserves estimates range from a few hundred to a few thousand years. Methane is considered a relatively clean burning gas.

There are a number of issues relating to the exploitation of hydrates. Firstly, they are not carbon neutral and pose the same problem as other fossil fuels in that respect. Secondly, there are the logistics of mining a resource several hundred meters under the surface of the ocean. Thirdly, methane is a greenhouse gas 20 times more powerful than carbon dioxide in its contribution to global warming. How much of it would leak into the atmosphere as part of the extraction process? Furthermore, it is believed that mining activities may destabilize the ocean's floor and cause landslides that may disturb hydrate deposits and result in the release of vast amounts of methane into the atmosphere.

Research is currently looking at carbon-neutral ways to exploit the resource. The methane removed would be replaced with CO_2 hydrates. Visit the following web site for more details: http://www.eee.columbia.edu/research-projects/sustainable_energy/Hydrates/index.html. Whether the new exploitation methods prove feasible remains to be seen. It would certainly much enhance the value of this resource.

We certainly cannot afford to add another 3,000 or 4,000 years' worth of greenhouse gases to the atmosphere. However, if carbon-neutral extraction methods are developed, methane hydrates do offer some hope in terms of significantly easing the transition to renewable energy. At this point, too many questions remain to be answered, one of them being whether we will ever be able to trust the industry to exploit the resource safely and report truthfully on leakages and disasters.

We have to come to terms with the fact that fossil fuels are on their way out. Many types of energy will be helpful in facilitating the changes that need to occur during the transition period. Carbon-neutral means of exploiting coal, petroleum, methane hydrates, etc. can help but are high risk and should only be considered if proven safe and as a means to bridging to a sustainable renewable energy future.

We also need a change in attitude: wiping out one resource after another to prop up our standards of living is an incredibly small-minded and selfish thing to do. We share the planet and its resources with all of the generations to come.

The New Global Warming Equation

Even with a successful Kyoto Accord, we will still continue to add huge amounts of greenhouse gases to the atmosphere. The strategy will not solve climate change problems, only slow down the process.

Global warming is ultimately a function of two factors: how much greenhouse gas is added to the atmosphere and how much is removed from it. We increase global warming not only by burning fossil fuels but also by deforesting the planet. We reduce it by cutting down emissions and growing vegetation.

Biomass has been stored over millennia not only as fossil fuels but also as live plant matter. As discussed earlier, when we harvest trees for one industry or another and reforest afterwards, we do not add net amounts of greenhouse gases to the atmosphere. Over the long term, carbon is re-stored into the biomass of the new trees. When we clear cut a forest without replanting, we reduce the total amount of plant matter on the planet. When this occurs, atmo-spheric greenhouse gas levels are increased exactly as if fossil fuels had been burnt because the carbon added to the atmosphere is never re-stored into new trees and forests.

If the total live biomass on the planet increases, it could compen-sate for some of the fossil fuels we are burning. The problem is that it is actually decreasing, not only adding to global warming itself but also reducing our capacity to absorb carbon from the atmosphere or compensate for fossil fuel use. The more trees there are, the faster carbon can be absorbed back into biomass. This capacity has been decreasing in the last few decades as deforestation has occurred in many countries, the clear-cutting in the Amazon rain-forest being only one of many examples.

With the discovery of methane hydrates, we now add a third component to the global warming equation: the indirect release of potent fossil greenhouse gases. As the earth warms up, we can expect that water will also see a rise in temperature. As this occurs, the massive beds of methane hydrates at the bottom of the oceans could begin to thaw out and release into the atmosphere large amounts of a greenhouse gas which is 20 times more potent than carbon dioxide. This will accelerate the greenhouse effect, which will result in the release of more seabed methane and the speeding up of global warming. This is a vicious circle that may make things happen much faster than anticipated. That was written in 2008. As I work on the 2010 edition of this book, the news headlines are reporting just that.

There are fears that global warming will likely result in the thawing of the permafrost in northern regions and the release of the millions of tons of carbon dioxide that are trapped in it. So, there might just be a fourth and significant component to the global warming equation.

How many other factors are yet to be discovered? In view of even only the last two components of the global warming equation, we have every reason to speed up the shift to renewable energies.

We should get used to the idea of droughts and increasingly disruptive weather patterns. The flooding of coastal areas will likely also occur much sooner than anticipated. Rebuilding New Orleans may turn out to be a major mistake. The fate of the Maldives is also very likely already sealed. The highest point of the chain of islands is about 2.3 meters (7.5 feet) above sea level, and most of its land-mass is only 1.5 meters (4.9 feet) higher than the surrounding waters.

4. The Resource Conservation Failure

The Rise of Consumerism

Technology grew by leaps and bounds through the 20th century. A large number of inventions have enabled us to produce all kinds of gadgets to make our lives easier and our leisure time more interesting and entertaining.

But, has that technology run out of control? Producing more and more means that we are using up more and more non-renewable resources. And, we are now over six billion consumers on the planet. According to Malcolm McIntosh (2000), a writer, broadcaster, and lecturer on corporate responsibility and sustainability, "Since the mid-20th century the world has consumed more resources than in all previous human history" (p. 47).

Minerals are limited in supply and do not belong to us alone but also to future generations. In but a few decades, we have used up several times our share. What is our plan for the next half century? Double that?

Today, we use up materials at a much faster rate than a few decades ago. As such, we might match the amount of resources used between 1950 and 2000 in only the first 20 years of the 21st century, or less for that matter.

Worse, the rate of resource depletion will only accelerate as developed countries only want more and more. In addition, China has a population of about 1.3 billion and has been exporting goods for a long time. However, not until recently has it seen enough income growth to support a significant amount of consumption. With its recent wave of trade and economic liberalization, things are changing fast.

India is not far behind in terms of population, and its economy is growing equally fast. Both are entering an era of consumerism. Together they represent about one third of the entire world population, currently estimated at 6.5 billion. The amount of resource depletion that will result from the growth of these two countries alone will simply be staggering. In the next 50 years, we are likely to use up three to five times the total amount of resources consumed in the second half of the last century.

The Corporate Solution

Corporations do not suffer or die like we do. As resources become scarce, they will continue to sell goods to us and keep making profits for their owners. In fact, the oil experience has shown that shortages often result in greater profits for them... and, of course, much higher prices, pain, and suffering for us.

What they do is use the cheapest, most economical resources first. For instance, the oil that is the least expensive to extract is pumped out first and used up. Then, the next cheapest source is used. Once it is exhausted, they move on again, so on and so forth.

As the price of resources goes up, substitutes that were more expensive or not as suitable become profitable and can be exploited. For example, natural gas, which is generally more difficult to handle than petroleum, could partially replace oil in transportation as the latter becomes scarce. When natural gas reserves suffer the same fate as petroleum, the market would again look for the next less suitable or pricier alternative, etc.

Problems With the Business Resource Model

The beauty of the above is that corporations will still make profits when the price of a liter of gasoline reaches $10.00 (about $40 a gallon). In the 1970s the giant oil corporations were criticized for price gouging. Three decades later, the headlines on CNN/Money read, "Big oil CEOs under fire in Congress. Lawmakers spar with execs from Exxon, Chevron over high prices, record profits, consumer pain" (Isidore, 2005, November 9). If you follow stock market news, you will notice that when the cost of a barrel of oil goes up, the price of petroleum industry stocks increases, signifying an expectation of greater profits.

Corporations will make profits selling us gasoline at $2 a liter. They will also do so when its cost reaches $20 a liter. We, however, will suffer a great deal. Resource depletion is not good for us, and the situation will be much worse for the generations that will follow us. The business model leads to one thing: the depletion of one resource after another.

Planning for the Future

Successful planning anticipates problems and fixes them before they arise. We need to plan ahead and conserve minerals to prevent a catastrophe from happening. Once non-renewable resources are gone, they are gone. Depletion is irreversible and will leave future generations with high resource costs, dysfunctional economic structures, and much lower standards of living. The stakes are high.

Actual estimates of reserves of different minerals vary not only from year to year but also according to technological developments, politics, and geopolitics. Recoverability and prices are also variables that make it difficult to determine accurately how long resources will last. The issue will be discussed in more details in the second book of this series. Suffice it to say that the only absolute in terms of non-renewable resources is that estimates are in terms of dozens of years for most minerals and a few hundreds for certain ones—not the thousands and millions of years that they will be needed for. In the long term, there is only one trend: reserves will decrease and prices will rise sharply.

Known oil reserves were extended as a result of a number of scientific discoveries, processing innovations, new exploration efforts, etc. Petroleum prices did increase more slowly on their account. However, most of these factors do not have a significant impact over an extended period of time although they did extend reserves temporarily.

The consensus among economists and experts with respect to oil is that its price will only continue to go up in the long term. Liberal estimates are that reserves will peak in 10 to 20 years. Some market analysts argue that they reached their highest levels around

2004-2005.

The Case of Energy

Energy looks like a poster child for the business resource exploitation model because of its substitutability. Although oil itself is not renewable and will run out eventually, it is highly substitutable. That is, when it and other fossil fuels are gone or get to be too expensive, we will shift to energies that are plentiful, unlimited, and renewable. That model only works because many of the long-term alternatives to petroleum and other fossil fuels are good and relatively inexpensive substitutes.

In fact, the whole transition to renewable energy has already begun. The world will eventually get by on hydro, wind, biomass, and solar power at a relatively low cost.

The business model does not work with other mineral resources, however. Its fundamental aspects are true: use the cheapest source first, and move on to the next cheapest after that. Science has also led to greater efficiencies. But...

The Case of Metals

One problem with the business resource model as it pertains to non-renewable resources is that energy is one of the few fields where the theory works. Would all other mineral resources have the same characteristics as energy, conservation would be much less of an issue. However, that is not the case.

The Substitution Argument

Steel, aluminum, and copper can be substituted for each other in many applications. They are mainstays of the modern world, being used everywhere in buildings, electrical infrastructure, and a variety of consumer goods. Although there is the possibility of substitution, these three metals are not renewable. They are all being used up simultaneously and would generally see their costs steadily increase as time goes by. If their depletion rates and costs go up in parallel to each other, then they are not true substitutes. That is, one could not replace the other in case of depletion.

For example, if 50 years from now reserves of iron have been exhausted, you would not be able to switch over to a plentiful supply of aluminum because it would also have been used at the same rate and be in shorter supply or near depletion by that time. We might even have already been considering switching from aluminum to steel as our supplies of the lighter metal ran low. If a metal is not renewable and is being depleted at a similar rate as another one, it would not solve a shortage problem and therefore not be a true substitute for it.

Furthermore, an excessively priced alternative—as would be the case if its reserves were highly depleted—would be useless. A true substitute needs to be both plentiful and reasonably priced at the time of substitution, i.e. not now but when the first resource is near depletion.

The Suitability Issue

Although metals may theoretically replace each other in many applications, they are not necessarily good alternatives. For example, gold, steel, and lead could probably all be used in electrical wiring but would be poor substitutes for a number of reasons. Gold would be extraordinarily expensive, steel would lack flexibility, and lead is toxic.

Most metals would actually be very poor substitutes for each other because of economic and suitability issues. Their use as such would create a dysfunctional society and be the result of the actions of very desperate people. Furthermore, this would probably occur at a point when society itself has already reached a state of economic crisis. Socio-economic dysfunctionality is likely to increase as resources are being exhausted.

To a large extent, substitution is a myth when talking about non-renewable resources. For us to live in a fantasy land with illusions of unlimited resources and substitution is dangerous. The reality is that, rather than jumping from one mineral to another as they are being exhausted, the world will see most metals depleted more or less concurrently. When their prices begin to rise sharply—just like oil—panicked and dysfunctional substitution will make little difference.

The Issue of Massive Use

There are no true substitutes for most common metals because they are all mainstays of the modern world and massively used. Alternatives would have to be available in enormous quantities at the time of substitution, which will not be the case. Furthermore, because of the massive use made of them, the substitutes would be quickly depleted, giving us but a very short reprieve, if any.

There are no real substitutes for many of the basic materials on which society's infrastructure is built. Their massive use underscores our fundamental dependency on them.

Once again, there will not be a substitution or jumping from one resource to another and another ad infinitum into the future. There will be a gradual increase in prices until that process starts to accelerate. By then, it is going to be too late. Panicked substitution will barely mitigate the problem and only last for a short time before reality comes crashing down on us. That process will likely begin around the middle of this century, that is, within your own lifetime. The issue is further discussed in the second book of this series.

The Issue of Resource Ownership

Another problem with the business model is the issue of price and ownership of non-renewable resources. These do not belong to us alone but to all generations the planet will see over its lifespan. The business model blindly skips over that part. It assumes that all these resources are ours to waste at will and with reckless disregard for anyone else coming after us. In economics, non-renewable resources are actually considered capital items—not on-going manufacturing inputs as they are treated at the moment. This has the strict implication that they should be used as sparingly as possible.

The Scientific Breakthrough Argument

Another major problem with the business resource model is the scientific breakthrough argument. One of the most powerful forces behind the fantasy world of unlimited non-renewable resources is the belief that science will solve all our problems. It has not in the past, and there is no evidence that it will do so in the future.

The business resource exploitation model is based on the assumption that future scientific discoveries will be made and that we can waste non-renewable resources in anticipation of that. By doing so, we seriously mortgage our children's future. Planning should be based on reality, not fantasies. Just a decade ago, most people thought that oil would last forever and remain cheap. Resource economics did not support that. The current state of science is that minerals (except energy) are not renewable and do not generally have true substitutes.

The Easy Science Issue

Science behaves to some extent like a resource. For example, in exploiting minerals, the most plentiful and easily accessible deposits are usually wiped out first. In several countries, including the U.S., many of the mines closest to population centers have already been depleted. As a result, resources are increasingly found further and further away and are more and more costly to process and bring to markets.

Science follows a similar pattern. Centuries ago, few things in the physical world were understood. Discoveries that appear to be insignificant today—the invention of the wheel, the mastering of fire—were major breakthroughs that changed the dynamics of entire societies. Today, we are much further ahead.

Scientific fields still in their infancy—the computer and information technologies, genetics, medicine, biotechnology—will continue to see lots of new and exciting developments. However, in the physical resource field, where knowledge is at a more mature stage, we should expect the breakthroughs to generally come less easily and less frequently as time goes by. In science like in other things, we have to make the difference between speculative illusion and reality.

The Five-Billion-Year Question

The earth was formed approximately five billion years ago. Its remaining life expectancy is about another five billion years. Ultimately, non-renewable resources should be managed in such a way as to last that long as they also belong to the generations that will live at that time.

In the last 50 years, we have used up as much of the earth's resources as have all the generations before that. In the same period of time, we have depleted maybe 25% of the known oil reserves. Experts estimate that in about 10 to 20 years these will have peaked and will begin to decline. In total, the bulk of world oil reserves will have lasted maybe 200 to 300 years. If the earth's lifespan had been 24 hours, our oil reserves would have lasted less than a second!

The reality we live in is not one of unlimited resources and infinite science. It is one where the physical capital (mineral supplies) on the planet is very limited, especially in terms of supporting large populations.

There are two very important distinctions to be made with respect to metals. One is that they do not have true substitutes like oil does. As such, you can expect much steeper price increases. The other is that if we wipe them out, there will not be a second chance because of their lack of substitutability.

The petroleum experience has shown us that as reserves decline, costs increase and commodities become the object of power. As time goes on, the process accelerates and price hikes turn into spikes. The summer of 2008 gave us a very brief glimpse of how quickly things can go bad. We would still be there today had it not been for the financial crisis. Other mineral resources will likely follow a similar pattern, with irreversible and devastating consequences.

Manganese Nodules: Panacea or Temptation?

First discovered in 1803, manganese nodules are potato-size nuggets of rocky material containing manganese, iron, and a number of base metals. They are found in many sites around the world, generally thousands of meters below the ocean's surface. They lie in large seabed deposits and in significant quantities.

They are seen as a potential source of ore for the future as reserves of surface metals become depleted. Manganese nodules could be a renewable resource as they are believed to be formed by bacteria depositing minerals from sea water onto their surface. They grow very slowly, at a rate of about 2 mm per 1,000,000 years. Their renewal speed is believed to depend on the amount of surface available to receive mineral particles. Mining them will reduce the total area for depositing and result in slower growth rates.

There are many issues with respect to their exploitation. Firstly, there are environmental concerns. There are also questions about our ability to extract minerals two to five kilometers below the ocean's surface. Their exploitation may turn out to be uneconomical or simply unfeasible. As manganese nodules only contain certain metals, they would not solve all our problems. Their excessively slow growth may mean that the new resource is to some extent depletable. Lastly, this may be our last frontier in terms of mineral reserves. We may want to preserve it for future generations and formally set it aside until we have reached a certain point in the future.

It is difficult to estimate how long existing surface resources will last at our current rate of use. Forecasts are from a few decades to a few hundred years, depending on the mineral. If we were to preserve enough resources for only 1% of the remaining lifespan of the earth, we would have to stretch what we have for another 50 million years!

We cannot even cope at this point with managing or preserving resources that are renewable. World species are dwindling and disappearing. We are totally impotent at preventing deforestation, be it in Nepal, India, or the Amazon basin. The cod fishery in Eastern Canada has all but been wiped out. Seabed resources are probably the only thing future generations will have left after we are done. The last thing we want to do at this point is to move into this last frontier. The solution to our problem does not consist in wiping out one resource after another. It lies in bringing ourselves under control.

Managing Resources for the Present and the Future

The earth has been around for five billion years, humans have been around for less than 10 million years, and civilization, for under 10,000 years. The planet has another five billion years to go. In 200 to 300 years, we will have practically wiped out petroleum resources on the planet! Other mineral resources have already seen their prices rise and are, in many cases, but a few decades behind oil.

So, what do we do?

5. The Silent Poisoning of the Earth

Over the last few decades, we have addressed some of the most obvious environmental problems. However, our efforts have only touched the surface, dealing mostly with only the worst crises and only once lives are at stake. Many less poisonous elements—but extremely damaging because of their pervasiveness—are slowly but surely accumulating in the environment.

This section provides a brief overview of some environmental contaminants. Its intent is not to cover the field in a comprehensive manner but rather to give some perspective to the pollution debate, show the extent of contamination, and support the case that the earth is slowly being poisoned. For those who may want more details, there are many works published on the subject, among others, Nadakavukaren's *Our Global Environment: A Health Perspective* (2000). Those already familiar with the topic should feel free to read selectively.

Historical Perspective on Contaminants

We have known for centuries that lead, mercury, and asbestos are health hazards. They are not new enemies but old foes. For example, author and lecturer in environmental health, Anne Nadakavukaren (2000) writes,

> Hippocrates described the symptoms of lead poisoning as early as 370 B.C.; mercury fumes in Roman mines in

> Spain made work there the equivalent of a death sen-
> tence to the unfortunate slaves receiving such an as-
> signment. (p. 225)

The 20th century saw the development of even more toxic substances. Scientific progress, rapid economic growth, and mass production spurred on the phenomenon. They introduced to the environment an entirely new array of compounds. Many of these are now part and parcel of our lives, found everywhere from the Arctic to the Antarctic, to the tissues of human adults and the unborn. They were and still are spewed out of smokestacks or flushed down our rivers on a daily basis. Others are accumulating at waste disposal sites and elsewhere.

Polychlorinated Biphenyls (PCBs)

First manufactured in 1929, PCBs are extremely stable in the environment and better known for their use in electrical transformers and capacitors. They found their way into the environment through electrical equipment catching fire, the burning of certain types of wastes, and illegal dumping into waterways by unscrupulous corporations trying to avoid disposal costs.

PCBs are toxic to several species at low concentrations and result in a variety of birth and health problems, including liver disease and cancer. Recent research points to their causing endocrine problems in humans and significant damage to developing embryos and fetuses.

PCBs have the ability to bioaccumulate, i.e. to concentrate up the food chain, from preys to predators and humans. According to Nadakavukaren (2000), in early research "virtually every tissue sample tested, from fish to birds to polar bears to animals living in deep sea trenches, contained detectable levels of PCBs" (p. 232).

PCB production and use in open systems were banned in the U.S. in 1976. However, the toxic compound is still legal in closed operations. As such, PCBs still pose a threat today. How much is left out there in warehouses and equipment in the custody of corporations? How much will eventually be leaked into the environment or get dumped illegally?

Dioxins

Dioxins (polychlorinated dibenzodioxins, PCDDs) are a large group of chemicals related to PCBs. They bioaccumulate and end up in the environment in a number of ways: as byproducts of some manufacturing processes, through the incineration of medical wastes and PVC plastics (polyvinyl chloride), via the smelting of metals, as a result of natural causes, etc.

Dioxins are believed to cause a number of health problems, including chloracne, developmental abnormalities, immune system interference, thyroid disorders, cancer, and diabetes.

Regulations have resulted in significantly lower levels of dioxins in the environment. They are often present in fatty tissues and foods like eggs, meat, fish, and dairy products. As a result, dioxins can be found in everybody, with industrialized countries being more affected. Breastfeeding is believed to increase the chemical's concentration in children's tissues (*Polychlorinated dibenzodioxins,* n.d.).

Asbestos

Asbestos' reputation as killer is well established. Nadakavukaren (2000) reports that 30% to 40% of the current and retired asbestos workers who have been exposed to large amounts of the mineral are expected to die of cancer (p. 243). Many others will suffer from asbestosis, a crippling lung disease.

Asbestos is a fibrous mineral found around the world. Its harmfulness was known to ancient Greeks and Romans, the mineral having been found to cause lung problems to slaves wearing clothes made with it. The fabric could be magically cleaned by simple exposure to fire (*Asbestos,* n.d.).

According to the U.S. Environmental Protection Agency (EPA), about 700,000 buildings (residential as well as commercial) in the U.S. contain some of it in friable form. The EPA further estimates that over 6,000,000 children and teachers may be exposed to fibers everyday in schools (Nadakavukaren, 2000, pp. 243, 246).

Lead, Mercury, Vinyl Chloride, Fire Retardants, and Jet Fuel

Both mercury and lead have been known for a long time as health and environmental hazards. Romans once lined their wine casks, cooking ware, and aqueducts with the latter. Lead poisoning can lead to mental retardation and death. Its main use today is in car batteries. Nadakavukaren (2000) reports that more than three million tons of it are mined every year and that "not surprisingly, lead is now found throughout the environment—in soils, water, air, and food" (p. 250).

Mercury, the quicksilver of ancient times, has been known and used for more than 2,500 years. It can damage the liver and kidneys and is believed to be responsible for a number of nervous system ailments. Mercury bioaccumulates and is found throughout the environment especially because of its ability to evaporate. It is present in many fish species and continues to be added to the environment from, among other things, the combustion of coal, the incineration of medical wastes, and the smelting of some ores.

Another significant environmental concern is vinyl chloride. It is a known carcinogen that is released into the air and ground water as the millions of tons of PVC plastics (polyvinyl chloride) we produce every year break down in the environment (Markowitz, 2002, pp. 9-10).

Brominated fire retardants are often sprayed on plastics to reduce their flammability. They are thyroid toxins which bioaccumulate and persist in the environment. A study by the Environmental Working Group (EWG) in the U.S. found that these chemicals were present in surprisingly high concentrations in all their samples of American women breast milk (Lunder and Sharp, 2003, September 23). More information on the study can be found at the EWG web site (www.ewg.org/).

In 2004, perchlorate—a rocket fuel ingredient linked to thyroid damage—was found to be present in cow milk in California and in the drinking water of almost half the states in the U.S. (*Rocket fuel*, 2004, June 22).

As seen in the few examples above, the earth is rapidly being poisoned. We already live in a chemical soup, one that not only pervades the environment but also permeates our bodies through and through. We are leaving our children and grandchildren a planet that is highly contaminated, and many toxic compounds are expected to continue to accumulate in the environment.

The Conspiracy of Silence

The responsibility for the current environmental crisis does not lie solely with some corporations. It extends beyond them. We keep silent while the earth is slowly being poisoned.

We have had some success with regulations and incentives, but they do cost money if not directly, then indirectly. Governments are not interested in making a costly commitment to the environment if it means that they will be voted out in the next election. We have to play our part in this.

In the last few decades of the 20th century, there was a lack of funding commitment. As a result, we failed to bring environmental issues to a head.

The Missing Link

There is a missing link between the green society that we need to achieve and the rallying cry of environmentalists. There is a reason why their efforts have remained largely fruitless, why we are destroying the environment for our children instead of preserving it, why we are decimating not only our share of resources but also those of hundreds of future generations.

Our failure is due in part to the lack of real commitment to the environment: money. It is also partly on account of the absence of an economically-viable strategy powerful enough to turn things around for the environment. What we have done so far has not worked. Regulations and had hoc funding is just not enough. We need new thinking.

We have to create an economic environment that will bridge the gap between theory and practice and reshape the current system into a mean lean green machine.

6. A Comprehensive Environmental Strategy

Commitment to the environment has picked up recently but still falls short of what we need. We have to shift gear, or we will leave behind a devastated world to future generations. We need fundamental change.

The only strategies powerful enough to turn things around for the environment are those targeting the very structure of economies. Patching up after disasters or dealing with symptoms has not worked so far and will likely continue to fail us in the future.

We need to use markets to our advantage and make them work for us. This could be achieved by changing the incentive structure of the marketplace in order to create an economic environment in which environmental practices would be more profitable than their opposites. The new structure would make green goods cheaper than others, prompting a shift in consumption patterns and industrial practices.

The First Principle

The first principle of a comprehensive environmental strategy is the recognition that a significant part of environmental contamination and degradation is directly related to the amount of non-renewable resources we dig out of the ground. Decreasing our consumption of these resources is key to both conserving them for future generations and reducing environmental degradation.

Many environmentalists campaign hard when it comes to global warming and pollution, but few call for action on the conservation of non-renewable metallic resources. Yet, these issues are intimately connected, one being a significant part of the solution to the other.

One of today's generally accepted principles in science is that *nothing is created nor destroyed*. Antoine Lavoisier (1743-1794) and Mikhail Lomonosov (1711-1765) are credited with its development and formulation. The *law of mass/matter conservation*, as it is known, pertains to the fact that things in the physical world are not destroyed when they burn or decompose. They are simply transformed into something else.

As a kid, you saw a log burn and assumed that the combustion process destroyed it almost completely, that most of its matter simply disappeared. In fact, it was only transformed. Some of it was released into the atmosphere as water vapor, gas, or smoke. Certain elements burned down to ashes, and the energy that was captured from the sun through photosynthesis was radiated back into the environment.

According to the Lomonosov-Lavoisier law, the tons of ore that we mine every year are not magically destroyed after we are finished with them. They end up in the environment. Some of what we extract from the ground becomes refuse at mining sites, and some is emitted into the air or discharged into toxic lagoons. Jared Diamond, an American evolutionary biologist and UCLA professor, describes the hardrock (metals) mining industry as "currently the leading toxic polluter in the U.S., responsible for nearly half of reported industrial pollution" (Diamond, 2005, p. 452).

The rest of what is extracted in the mining process is made into goods that eventually end up in landfill sites. But the story does not end there. As Annegrete Bruvoll, the Head of Research at the Unit for Energy and Environmental Economics (Statistics Norway) reports, "End treatment, i.e. waste disposal and incineration, results in emissions of toxic pollutants and greenhouse gases, and seepage from waste disposal sites pollutes ground water and watercourses" (Bruvoll, 1998, p. 16).

The total amount of minerals we extract every year is exactly equal to the amount of wastes we generate (refuse, emissions, fluid discharges, garbage, gases and solids from incineration, etc.). One ton of extracted minerals today means one ton of waste added to the environment tomorrow. We dig out ore annually by the millions of tons. Therefore, by the same millions of tons, we create wastes and pollutants every year!

Reusing and recycling are the closest thing we have come to so far in terms of a solution to the problem. Unfortunately, our best efforts only minimally delay environmental degradation. Most things can only be reused or recycled so many times.

In the long term, our two most prominent achievements, reusing and recycling, will only give us a bit more time as everything will sooner or later end up in the environment. Although their impact is very significant in the short term, they cannot bring change on the scale needed and are only part of the solution.

Conservation, on the other hand, achieves two goals at the same time. It preserves resources for future generations and massively reduces wastes and pollution: one ton for every ton of minerals not mined or extracted.

The Second Principle

The second principle of a comprehensive environmental strategy is that cutting down on the extraction of minerals would not only preserve resources and massively reduce pollution but also achieve a third goal: decrease our use of the battery of invisible toxic compounds produced in the processing of ores and manufacturing of goods.

Many of the toxic chemicals that are released into the environment today result from the transformation of the minerals we mine and their manufacturing into finished goods. A range of compounds— *intermediate chemicals*—are produced for those purposes or are by-products of manufacturing. This is also true of certain toxic metals. As such, they often end up in lagoons, waterways, and the atmosphere just like intermediate chemicals.

For example, mercury is at times used in the extraction of gold. It is also contained in the ores of other minerals, the smelting of which releases the volatile metal into the atmosphere. The burning of coal for electricity or other purposes also results in the emission of mercury into the air in addition to all the other pollutants its combustion produces.

By virtue of the fact that nothing is created nor destroyed, intermediate chemicals and some toxic metals also end up in the environment, in their original forms or as byproducts.

The Third Principle

The third principle of a comprehensive environmental strategy is that we have to reduce world population. Even the greenest measures and initiatives will not be able to compensate for the devastation that billions of people can wreak upon the earth. The greater it is, the more resources are consumed at any given point in time.

For example, if the total number of people on the planet were reduced by 30% and everything else remained the same, we would theoretically consume 30% fewer goods and use up approximately 30% fewer resources. We would need to extract 30% less mineral, burn 30% less fossil fuels. We would create 30% less waste, pollution, and degradation. Industrialized countries consume far more than their fair share of resources. Reducing their population would have a much larger impact on mineral reserves and the health of the planet.

Today's efforts towards the environment only skim the surface of problems and have no hope of ever catching up to the current rate of destruction. There is just too much being dumped into the environment, from the billions of tons of minerals extracted every year, to the millions of gallons of intermediate chemicals, to the multitudes of soaps, detergents, solvents, and cosmetics flushed down rivers on a daily basis.

Funding for a Comprehensive Environmental Strategy

Taxation

Funding has always been the stumbling block of the environmental movement. Fixing problems costs money. This is why our efforts are still only skimming the surface of problems.

People often complain of giving too much of their hard-earned cash to governments. Doubling taxes to raise the trillions needed for the environment is not going to happen. The countries that do tax for the environment have only been able to raise minimal amounts for that purpose. There is no particular support, either on the part of politicians or voters, for any large amount of additional taxation. Even the most charismatic environmentalists already have enormous difficulties raising money to support their own work.

As it stands, additional taxation can only have a minimal impact for the environment because of its hopelessly insufficient fundraising capability.

Subsidies and Tax Breaks

When governments give grants or tax breaks to corporations, they have to make up for the shortfall in revenue through additional income tax, cuts in services, or other similar actions. Subsidies and tax breaks are not as free as many would think; taxpayers ultimately foot the bill. As such, these approaches to resolving environmental problems are nearly as limited as additional taxation is.

Regulations

Regulations have scored a number of victories for the environment, for example, outlawing PCBs in open systems and chlorofluorocarbons (CFCs), the chemicals responsible for the depletion of the ozone layer of the atmosphere. However, they are also ultimately limited by the availability of funds. PCBs, for example, are still widely used today in closed systems. CFCs will only be totally phased out in 2030 despite being the object of significant restrictions and bans since the late 1980s. Why is that?

Regulations do cost money either in the form of higher prices for substitutes or because of conversion expenses. These are usually passed on to consumers. Once again we pay the bill, this time in the form of more expensive retail products. This is why regulations are often much weaker than needed, assuming they make it to the political agenda and are enforced.

Short of people dying, there is a limited amount of social commitment to seriously address environmental problems and even less with respect to long-term preemptive planning. Regulations will remain an important tool for the environment, but alone they have no hope of ever solving our problems. We may be willing to pay a little more for greener alternatives but not what would be required for an environmental turnaround or revolution.

Policies for the Future

Resource conservation has generally failed to make it to the social and political agendas. Most often, the only resources targeted with our efforts towards sustainability are forestry and fisheries. This is no great achievement as both are actually renewable. Despite that fact, we still managed to fail. The collapse of the cod fishery on the Grand Banks on the eastern coast of Canada and the U.S. is but a striking example of our lack of resolve.

We have made no significant progress on the issue of non-renewable resources, which is by far much more critical. The world hopes that OPEC increase production to lower gasoline prices, not decrease it to save reserves. Nothing is done to conserve metals. On the contrary, calls are being made to increase production in order to create jobs.

Sustainable development runs over and over again into the same brick wall. The very incentive structure of the system in which we live directly pushes us to deplete non-renewable resources to keep prices down or create jobs. Similarly, it makes it cheaper to pollute, use toxic compounds in the manufacturing of goods, and flush hazardous waste into waterways than opt for cleaner alternatives.

Today's reality is that the more we pollute and decimate resources, and the less we leave for our children, the greater our economic growth. Two things are needed to address environmental problems. Firstly, the solutions have to be on the scale of the problems. Secondly, the current economic structure has to change. No matter how hard we fight for the environment, if capital runs counter to our efforts, little will be achieved. How do we change that? Let us first look at what has been proposed before.

Hitting the Jackpot

The idea of environmental taxes is not new. In 1989, David Pearce produced for the Department of the Environment in Britain a document which recommended a comprehensive green taxation program in the UK (Dryzek, 2005, p. 130-132). The U.S., Canada, and many European countries have already implemented some form of environmental levies, albeit very minimally. The global warming debate has also led to talks of carbon taxes in several countries.

Could trillions of dollars in additional taxation for the environment solve our problems? Probably. The scale would be appropriate, but the idea would not be politically viable. Could we find trillions' worth of spare change within the existing taxation system? The answer is also very likely *no*. But...

Many countries have deterrence or punitive taxes such as levies on cigarettes and alcohol. These are meant to do two things: deter usage and generate income for the government. They are *dual-purpose* levies. When governments tax income, they bring in revenue but nothing is specifically being deterred. Because this is a *single-purpose* type of levy, society derives from it only one benefit when it could have had two.

   Shifting taxation to the environment would double the social dividends we get from our tax dollars. This would create synergies that could achieve a lot while not costing us a cent. There might not be trillions of dollars of spare change available within the current taxation system, but there certainly is that amount of wasted incentive or social benefit.

   Considering that Americans and Canadians alone are paying about US$ 2 trillion annually in income tax, the benefits of such a strategy could be massive and result in a complete turnaround for the environment in as little as a decade. Let's take a closer look.

A New Budget for the Environment

There is no doubt that levies have to be collected to fill governments' coffers and fund social services, healthcare, education, roads, the police, the army, etc. But, who is to say that income has to be taxed in order to raise the money for the above? Canadian taxation represents over 30% of national income. In the U.S., the figure is somewhat lower but remains a massive component of the country's revenue. Those represent staggering amounts.

   Let me illustrate with a concrete example what a shift in taxation could do. For the purpose of the exercise, $1 of deterrence is considered $1 of benefits although it is not clear how much direct savings the deterrence would translate into for society. It could be less but also more.

Income tax is a single-purpose levy designed to raise revenue for governments: $100 of it produces $100 worth of services. With deterrence taxes, governments collect revenues to provide services as well as deter the use of products which are generally considered unhealthy: $100 of those provides not only $100 worth of services but also $100 of deterrence. In the case of cigarettes, the latter could translate, for example, into lower healthcare costs from decreased cancer rates.

In North America, when you tax income—as opposed to goods like cigarettes—you get US$ 2 trillion into the government coffers, but you lose another US$ 2 trillion worth of deterrence effect. That alone could radically turn things around for the environment without costing the taxpayer a cent, i.e. without additional taxation!

Most environmental issues are ones of deterrence. We want to deter the emission of greenhouse gases, the use of pollutants, and the depletion of non-renewable resources. Shifting the single-purpose tax on income to an environmental deterrence system would double the social benefits we get from every dollar of tax paid without increasing taxation. The US$ 2 trillion of deterrence that this would generate for North America alone would be a massive engine of change for the environment.

Our lifestyles would change, but that is unavoidable. As expressed earlier, the transition to a greener world will occur one way or the other, either destructively or constructively. However, with the strategy proposed in this book, our standard of living would not go down. Disposable incomes would essentially remain the same as the total amount of taxes paid before and after the shift to dual taxation would essentially be equal.

We already know that there is no social commitment to any significant increase in taxation and that current spending for the environment is far from adequate. The shift to deterrence taxation would allow us to get a second bang for free from the tax system. If you currently pay $10,000 in income tax, you will still be paying about the same amount under a dual taxation system. However, a significant portion of it would then come from deterrence levies, for example on fossil fuels, contaminants, and non-renewable resources, instead of income. $10,000 worth of revenue would still be generated for the government, but the punitive effect would reduce greenhouse gases and contaminants, and foster resource conservation without your paying any more taxes than before.

In 2000, Canada collected on incomes about CAD$ 127 billion (Revenue Canada, 2002). Americans paid US$ 2,098 billion in taxes in total during the same year. Assuming that the currencies are at par (CAD$ 1.00 = US$ 1.00), this would amount to about US$ 2,225 billion (or $2.225 trillion).

In comparison, the U.S. gross national product (GNP)—a measure of the total value of goods and services produced by a country—for 2000 was about US$ 10 trillion. The actual annual environmental deterrence budget—the *green firepower*—for the U.S. alone would be about 20% of its entire national production.

Internationally, the story would be similar. The rest of the developed world would have a comparable share of its total production available for environmental deterrence. Developing countries, which have been able to afford environmental standards only with difficulty up until now, would have less but still enormous amounts of deterrence at their disposals. By any standards, we are talking about massive and unprecedented firepower for the environment in *both* rich and poor countries.

The GNP is very effective in giving us a good idea of the actual size of the new potential environmental deterrence budget. But, a much more interesting comparison is with the American military budget. It stood at US$ 379 billion for 2003 (Council for a Livable World), in the midst of the Afghan and Iraq Wars. At $2,225 billion, the new budget for the environment would not only beat but actually dwarf the war chest of the most powerful military on the planet.

With this kind of budget, an environmental revolution becomes more than a mere possibility. It is right in our hands, just waiting to happen.

7. Premises of Large-Scale Environmental Change

Turning things around for the environment would imply change on a large-scale. At first sight, it may seem impossible to achieve. However, if certain principles are respected, it can happen. There is a science to massive change.

The First Premise: Massive Scale

The first premise of a meaningful environmental strategy is that needs are massive. Therefore, the solution should be of appropriate scale, powerful enough to address all issues. There is an urgency to deal with global warming and environmental degradation. There is also a dire need to reverse the current trend of rapid resource depletion so that future generations can have enough for themselves. The sad reality is that the earth does not come with a warranty stating that because its life expectancy is five billion years, its resources should last that long.

The small step by small step incremental solutions proposed so far will not do the job.

The Second Premise: Strong Political Support

The second premise is that environmental change on the scale needed would require massive voter support. For any solution to be implemented, it must first make it to the political agenda and receive support. To that purpose, the strategy explored in this book is revenue neutral, making it as voter friendly and as politically viable as possible given the ambitious task at hand.

Because society's support for the environment is ultimately limited, large-scale change would have to be achieved with virtually no additional amount of social commitment. This brings us to the next premise.

The Third Premise: Minimal Social Commitment

The third premise is that to be voter friendly and get massive support, strategies for large-scale environmental change have to be able to be implemented essentially without additional social commitment.

Current solutions always run into the same obstacle: voters do not want any additional taxes, at least not on the scale required. Designing strategies based on a massive increase in support is unrealistic. We need solutions that are able to address most environmental problems with roughly the same amount of social commitment as we have today. The revenue-neutral approach proposed in this book does provide for this.

The Fourth Premise: Implementability

The fourth premise is that large-scale environmental change calls for practical and implementable solutions. Remember Keynes' famous quote, "Foul is useful and fair is not"? Rather than an attempt at cynicism, it was an acknowledgment of the basic reality of economics. Theories and principles are not very useful if they are not applicable. If he had spoken about the environment, he might have said, "What does not make it to the political agenda and sway the voter has very little use."

Some of the money spent on research every year does not benefit society for a number of reasons. Studies might not result in practical applications, they might not have been brought to the general public's attention, or the solutions proposed are just too unfriendly to voters to be politically viable. Of course, not all research is meant to produce solutions. Some types are supportive in nature.

The 21st Century Environmental Revolution was written strictly from an implementability point of view. What is implementable and politically viable is useful; what is not, is not. The strategy described in the next chapters is intended to be as practical and voter friendly as it can be, considering the fact that we are contemplating massive change.

The Fifth Premise: Market-Efficient Mechanisms

The fifth premise is that, to maximize implementability and political support—especially given the magnitude of change being considered—environmental solutions should make use of the most efficient mechanisms available. Let's review the options at our disposition.

Current research divides approaches into two main categories, regulations (*command-and-control*) and taxation. Regulations can be very effective in specific cases and have their place on the environmental agenda. However, they generally lack efficiency on several counts.

Ian Bailey (Senior Lecturer in the School of Geography department, University of Plymouth, UK) reports that they are viewed as economically inefficient because they are imposed uniformly on polluters who have different abilities to adapt to better practices (Bailey, 2002, p. 235). For example, big corporations can more easily invest in cleaner technologies than smaller businesses.

Annegrete Bruvoll (1998) further argues:

> The required control costs under regulations can be considerable and the administrative costs are usually higher than under taxes. Also, price incentives offer free cost information, while mandated recycling is often accompanied by expensive education and motivation programs. (p. 19)

In other words, regulations often have a number of costs associated with them—administrative, implementational, educational, etc. The same author also reports that studies had consistently shown regulatory programs to be poorly designed and have costs exceeding benefits.

Another problem is that, unlike taxation—which promotes

continued improvement—regulations are not *dynamically* efficient. That is, they offer little incentive for polluters to exceed minimum requirements and do better year after year. In addition, without international agreements to back them up, regulations are further limited as they tend to impede the competitiveness of countries which implement them.

For these reasons, the focus of policy has recently been shifting to market-based solutions. These new approaches are increasingly viewed as the best ways to address environmental issues. Bailey (2002) states that "both theoretical and empirical studies suggest that market-based mechanisms remain the most efficient method for pursuing many environmental goals, so that countries pursuing such strategies generally perform better economically than those that rely on legislative standards" (p. 237).

Note that the Kyoto Accord sets caps or limits on emissions. As such, it is a regulatory system, not a market-based solution. The *cap* part of cap-and-trade refers to its targets. The *trade* component refers to an exchange system that would need to be set up and that would make use of markets. However, it is only a flexibility component allowing companies that find it cheaper to pollute to continue to do so and buy emission credits instead.

As demonstrated further down, the trading system itself would not lead to a reduction of greenhouse gases. Shocked? As such, it is not an environmental strategy per se. What would bring down carbon emissions are the caps negotiated under the Kyoto Accord. The system does not help businesses or countries—whether through information, funding, research, strategic planning, or technology— to reduce carbon emissions. Under a cap-and-trade initiative, it is really up to the industry to figure out how the Kyoto Accord target are going to be achieved. Such an approach does not take too much thinking on the part of governments!

Market-Based Options

Bailey (2002) describes three main market-based options, *incentive taxes*, *cost-covering charges*, and *revenue-raising taxes*. Incentive taxes are designed to increase the costs of polluting in order to bring about positive environmental behaviors. For example, toxic chemical compounds or fossil fuels can be taxed to reduce their use.

Cost-covering charges are usually levied from users of a particular resource and used to manage it or mitigate its degradation. For example, entrance fees to national parks may be used for their maintenance. Pollutants might be taxed to raise funds to undo the damage they cause.

There are two types of revenue-raising taxes. The first, *additional taxation,* involves for example environmental charges on plastic soft drink bottles or residential tipping fees on garbage. The revenue generated through these go directly to general government tax coffers rather than being targeted for specific purposes. They are additional charges to consumers and businesses (which pass those on to us) just like the first two types of market-based approaches discussed above.

As such, all three options are limited by the current level of social commitment—nobody wants to face new user fees or pay more taxes, at least not on the scale required—and are therefore bound to only skimming the surface of environmental problems. This brings us to the last option.

The second type of revenue-raising options is revenue-neutral taxation. Environmental levies are collected and offset by lowering other taxes, for example, on income or retail sales. They do not increase the general level of taxation for people in a country. They are only a shift from one form of levies to another.

For example, if a government were to raise $1 billion in new environmental taxes, it would decrease income or retail taxes for taxpayers by the same amount. A government would take with one hand and give back with the other. In other words, because environmental taxation is offset by a drop in other taxes, it would be revenue neutral. The levies would result in a deterrence of unenvironmental behaviors—progressively making economies greener—without increasing total taxation for individuals.

This essentially gives us not only an efficient mechanism but also a politically-viable one. It is the basis for structural changes that would be powerful enough to bring about an environmental revolution.

The Sixth Premise: Focus on Resource Conservation

The sixth premise is that resource conservation should be at the core of environmental policy as it attacks the problem at the source through prevention.

As discussed earlier, a ton of mineral not extracted represents one less ton of garbage produced and a reduction of intermediate chemical use and environmental degradation from waste disposal. In addition, making conservation a primary focus of policy forces us to address the issue of non-renewable resource depletion. Current economic planning leads us to give little or no attention to preserving resources for future generations. In fact, it does the very opposite.

Another important reason for putting conservation at the core of environmental policy is that pollution and contamination are in most cases easier and cheaper to prevent than clean up. To what extent can gases from industrial processes or the incineration of garbage be cleaned up? Bruvoll (1998) writes, "Most [waste] materials are dispersed and hard to collect after use at the emission stage, and designing and operating a tax system that tracks all environmental effects is out of reach" (p. 16).

The Seventh Premise: Landfill Sites Are Limited

The seventh premise is that there is a limited supply of suitable sites to dispose of our garbage. Because of the toxicity of some materials and concerns about seepage of contaminants into water tables, not all locations are geologically suitable for waste disposal.

For economic efficiency, landfill sites are often located in close proximity to towns, threatening the contamination of drinking water sources. Current recycling efforts only partly reduce the continued demand for landfill space. They do not eliminate it.

8. Structural Change

The Current Incentive Structure

Under normal circumstances, you would expect governments, and certainly more so an economic system, to reward good behaviors and deter bad ones. Not doing so would lead to a very dysfunctional society and have harmful consequences. This is exactly what is happening with the environment.

The way the current economic system is set up makes it more profitable to consume carbon-intensive fossil fuels than renewable energy, to use cheaper toxic compounds in manufacturing processes, and to mine greater quantities of metals as it creates jobs and increases profits. In other words, current economics actually rewards the production of greenhouse gases and the wanton destruction of resources and the environment. Why is it so surprising that the planet is in a total mess? It is just the direct result of the current incentive structure of the economy.

The market system on which the entire world economy currently depends is configured into a destructive mode. It is setup to do exactly what it is doing now!

The Green Economic Environment (GEE)

The only approach capable of bringing change on the scale that is needed to turn things around for the environment would have to involve the economic system. One way to do this is to change the incentive structure that drives it, and the easiest and most obvious means to make that happen is to modify taxation to create a *green economic environment* in which non-polluting and conservational behaviors would be rewarded and their opposites, deterred.

This could easily be achieved by shifting the income and retail taxes that governments collect to environmental levies on non-renewable resources, pollutants, and fossil fuels. The changes would be done in such a way as to not increase overall taxation, hence, be revenue neutral. This strategy will be referred to as the Green Economic Environment (GEE). For the sake of clarity, the acronym "GEE" will apply specifically to the approach as opposed to *a* or *any* green economic environment.

A Brief Overview of the GEE

Our current taxation system can be characterized as a politically-driven shotgun approach to filling the tax coffers. It is dysfunctional economically and disastrous for the environment. At this point, not much thought is being given to the fact that taxation is the very incentive structure that shapes our lives and the economy of our countries.

The lack of emphasis on the environmental component of taxation leads to behaviors that make us pollute the planet and deplete and waste non-renewable resources as if there were no tomorrow. Current environmental problems are in large part caused by the improper structuring of the economic system in which we live.

That can be changed by altering the tax incentive structure of economies to reward environmental and conservational behaviors and punish their opposites. Under such a system, a significant part of government revenues would come from taxes on non-renewable resources, contaminants, and fossil fuels.

The GEE strategy proposed in this book would be revenue neutral. As such, a government would collect the same total amount of taxes after its implementation, but it would come from different sources. For example, if 60% of taxes came from income, 38% from retail sales, and 2% from environmental levies, the implementation of the GEE could result in a new distribution looking like this: 40% coming from income, 10% from sales (a total of 48% reduction for the two), and 50% from the environment (a 48% increase).

The new incentive structure would create a green economic environment and a new consumer landscape in which the price of non-renewable resources, pollutants, and all the products made with them would increase. The goods made with renewable and non-toxic materials would become cheaper and gain a competitive advantage.

As a result, consumption would naturally and effortlessly shift to greener products and sectors of the economy. Good citizenship would be rewarded rather than punished as it is now, and the destruction of the planet would be deterred rather than encouraged as is currently the case. The GEE would create a *greening effect* that would conserve resources, reduce contaminants, and decrease greenhouse gas emissions.

As a general rule, the GEE would involve taxing non-renewable resources—such as metals—at the source to promote conservation and create natural markets for recyclables. These would replace or make profitable the municipal recycling programs that are currently funded by taxpayers. Simple higher input costs would be more easily understood and managed by businesses compared to a haphazard collection of regulations.

Substances that are damaging to the environment (toxic compounds, carcinogens, etc.) would be taxed, whether they are inputs in manufacturing processes or consumer goods.

Oil and other fossil fuels would have levies assessed not only to decrease usage and greenhouse gases but also to create a stable and profitable renewable energy sector. This would be the simple mechanism by which countries would reduce their carbon emissions and reach their Kyoto targets. It could entirely replace the more complex and costly cap-and-trade system although both strategies could work side-by-side.

Regulations would be used in the packaging industry. In conjunction with GEE taxes, they would create natural markets for container reuse. This would serve to replace most of the existing refundable-deposit systems and result in efficiencies and a drop in costs for the consumer.

Of course, all of these new taxes would be compensated for by a drop in existing ones on income and retail sales since the system is revenue neutral.

The shift to a green economic environment would be fully scalable. That is, the GEE could be implemented as fast or as slowly as people desire. This would enable countries to make the transition to the new system as quickly as possible without creating economic havoc or unnecessary human suffering. Scalability would allow governments to start with lower taxation levels—giving consumers and the industry time to adjust—and progressively increase them over time. It would also make it possible for countries to give different weights to each of the components of the GEE according to their own goals and political imperatives.

For companies, the GEE environment would make it more profitable to preserve resources than waste them. The opposite would be true for the use of toxic materials and contaminants in the production process and the emission of greenhouse gases.

High-Efficiency Systems

Many environmental policies are inefficient and expensive to administer. They are cumbersome to the business sector as they require that staff keep up to date on compliance and regulations. It is usually much easier for companies to make decisions based on the price of the resources needed to manufacture a product (the cost of inputs) than have to deal with a maze of government rules. It is a language readily understood and simpler for planning and decision-making. It is already part of normal business practices and expertise. This is the language that the GEE also speaks.

Environmental taxation would raise the cost of an input in a predictable way. Businesses would make decisions on that simple parameter. Current environmental policy is cumbersome to governments because regulations need to be monitored and enforced, often necessitating additional costs and bureaucracy.

A second aspect of efficiency has to do with the total number of tax collection points. That is, all of the individuals and businesses that governments collect from. Removing part of the taxes people pay on income would eliminate the need for many individuals to submit tax returns or, at least, would make those much simpler. That could mean thousands fewer or easier filings to be processed.

However, much more interesting is the effect on retail levies.

Many governments currently collect retail taxes from just about every single business selling goods or providing services to consumers. This represents thousands of corporate taxpayers and means a lot of unnecessary bureaucracy.

As you go up the production chain—from primary resource extractors, to input producers, to manufacturers, to retailers—the number of individuals from whom governments have to collect taxes grows. For example, a copper mining company might sell its products to 10 part makers. Each of these in turn might sell parts to 10 manufacturers. At this level, 100 businesses would have to be collected from. Each of them might sell to 10 retailers. Collecting taxes at that level would then involve 1000 companies. In other words, the higher up you collect, the more costly it is not only for governments but also for businesses and the economy as a whole.

The GEE would collect taxes mostly at primary levels, therefore minimizing the number of players involved. Furthermore, it would do so only on target items: non-renewable resources, contaminants, fossil fuels, etc. Green sectors would generally see taxes disappear from their products.

Dual-Level Planning

There are two main types of planning in the environmental arena, the general (or macro) and intricate (or micro). Governments are responsible for planning at the general level. This is where they state what they want and establish plans of action to achieve certain goals. The difficulty with many environmental issues is that they are very complex to manage at the micro level. Problems are often just too intricate in the day-to-day reality of markets and business operations even for the best specialists in governments.

It is not really a matter of failure. For example, even with today's technology, we still cannot predict the weather with any degree of accuracy for more than a few days ahead of time. Currently, government regulations and programs are the main planning mechanisms used to address environmental issues. They fail to do the job. Furthermore, when they do not work, the taxpayer ends up footing the bill. A successful green strategy would need to have appropriate expertise at the micro level while retaining the ability to operate within general government guidelines.

The market has a multitude of agents—business people—each

an expert in a given field. Nobody else can better estimate the effect of higher input costs on the final price of a product or the bottom line. All planning on their part is based on familiar market decisions and comes free of charge to the taxpayer. The private sector represents a massive wealth of expertise at the micro level. If unqualified for planning at the general level, the business community possesses the detailed knowledge necessary to make the best economic decisions in regard to their own markets. Governments cannot hope to emulate or even approach that kind of expertise.

The GEE would rely on governments for general directions and on the market for decisions at the micro level. Setting tax levels would enable countries to guide the economy where they want it to go and allow market expertise to make things happen in the most cost-effective way possible. This would provide perhaps the best combination of planning and a much more powerful and effective strategy for the environment.

Business-Friendly Green Policy

The GEE would reduce the regulation and tax collection burden on industry. It would also provide for much simpler decision-making for businesses. The fact that the system would be highly flexible, fully scalable, and comprised of several components that could be implemented independently would allow for a smoother implementation for everybody.

New Approaches to World Development

A worldwide implementation of the GEE would give rise to several new mechanisms for world development. Since the 1970s, the developing world has been hit by several oil crises, preventing many countries from gaining significant ground on poverty. As renewable energies are diversified and can be produced nationally, developing countries would benefit from the GEE in terms of balance sheet and job creation.

Global resource conservation would increase the international price of most metals. Developing countries would get better prices for their commodities and could be allowed to have lower GEE tax levels to increase their international competitiveness and gain market shares, boosting their own economy and promoting their industrial development.

Environmental regulations have always been an issue of contention in international markets. The developing world has historically been perceived as having an unfair competitive advantage over developed countries because of lower environmental standards. The world cannot afford a race to the bottom.

Comprehensive GEE-based international agreements could raise standards across the board to benefit the entire world community while not affecting competitiveness. Conversely, a unilateral increase of environmental standards in the developed world could be used as a means to increase the competitive advantage of developing countries. The former would benefit from a cleaner environment while the latter would get an economic boost from a greater share of markets.

In more ways than one, the GEE would provide a simple and effective framework for international development.

Tax Fraud

To some extent, most social and economic systems are prone to fraud. The fact that the GEE would be levied at the manufacturer level would significantly decrease the number of collection points. As such, the system would be not only more efficient but also less prone to fraud than current widespread retail taxes.

Black market or undeclared work is a problem around the world. The bill for the *underground economy* can add up to hundreds of millions of dollars every year in a medium-size country. In Canada, for example, people do not pay income tax on their first $10,000 of income (the basic deduction). As such, black-market fraud up to that amount is mostly meaningless; no one pays income tax on it anyway. The GEE could shift up to the first tier of income tax ($30,000 in Canada) over to the environment. This would mean that $40,000 dollars of a person's income would no longer be taxable and subject to fraud. This could have a significant impact on government revenue and result in lower overall taxes.

A New Future

The GEE is a strategy that could bring about the large-scale change that is needed for the environment. The new incentive structure would create an economic environment that would turn a highly destructive market system into a powerful force for positive change.

The strategy would take advantage of the dual effect of levies— one that is wasted by income and indiscriminate retail sales taxes. With no overall increase in taxation, it would transform the economic system in which we live into a mean green machine and make environmental change possible to occur at a speed so far unanticipated and not even dreamed of.

The implementation of the GEE would change the incentive structure of capitalism from a system that essentially destroys the planet and punishes people for working (income tax) and consuming (sales taxes) to one that rewards them for doing the right thing and promotes resource conservation and environmental responsibility. The new incentive structure would foster this in both the industry and consumers.

The GEE would also be highly efficient. Its administration would be much less costly than our current patchwork of regulations as it would make use of simple and well-understood systems. It would speak a universal language, one understood by workers, business people, consumers, and voters: money. Once the new rules are set, societies would progressively refashion themselves around green economics and environmental lifestyles.

The scalability of the GEE would allow for choice in speed of implementation and help soften its impact on specific economic sectors. Its various components could be implemented together or independently as desired.

This structural approach to environmental issues would not be anti-business or anti-economic. It would work in concert with the main components of the current market system. Overall, companies would not be stuck with the costs of this green transformation anymore than taxpayers. On the contrary, the GEE would make it profitable for companies to invest in green practices and environmental research and development (ER&D).

Many business opportunities would arise following the implementation of this new tax structure. The current catering to green market niches would become a stampede for opportunities in an exploding new economic frontier. Countries and corporations at the forefront of ER&D would find themselves in a race to capture the new 21st century world markets. Wasters, polluters, and those who do not care for the environment or are unable to adapt would only fall behind and see the fate of the dinosaurs.

The GEE would give rise to a new overall mechanism for world development.

General Issues Relating to the GEE

Governments collect taxes from a number of sources, for example, income, retail sales, the environment (levies), and profits. In addition, specific products such as cigarettes and alcohol are often taxed for deterrence purposes.

For the sake of simplicity, the following discussion will concern mainly the first three types. Of those, income is probably the main source of revenue for most governments, with sales coming in second and environmental taxation being minimal. The GEE would progressively shift the tax burden from income and sales to the environment.

Keeping Taxation Progressive

Most Western countries have a progressive taxation system. That is, people contribute according to their ability to pay. For example, in Canada the tax rate on the first $30,000 of net income is about 30%. It is the first bracket. If that were the only one, we would have a flat tax system. That is, everybody would pay the same rate, no matter the income level. This first bracket is actually a flat component in a progressive tax system.

As one's net income rises above that, a higher rate is paid. For example, Canadians are taxed at 45% in the second $30,000 bracket. For the next two tiers above that, they pay respectively 55% and 60%. In order to keep taxation progressive under a GEE system, only the flat portion of income tax would be suitable for a shift to the environment. In Canada, that would be the first bracket, or the tax on the first $30,000 of net income after the basic deduction.

In addition, we also have to be careful with respect to how we go about eliminating that bracket. Simply increasing the basic deduction (about $10,000 in Canada at the moment) to reduce the total taxable income would be regressive because it would lower taxable incomes at the top where people pay the highest tax rate. That would be regressively cutting back taxes.

If a government chose to increase the basic personal deduction as a means of implementing a GEE system, the tax bracket thresholds would need to be adjusted lower by the same amount to avoid changing the existing distribution of incomes. Alternatively, a government may simply set the tax rate in the first bracket to 0%.

The GEE would imply a large reduction in income tax for everybody. Authorities have to make sure that it is done in a way that maintains the status quo with respect to progressiveness.

Most countries have retail sales taxes. This is what people pay on top of the purchase price of items bought in stores. This type of sales levy is not progressive as the same rate is charged equally to everybody. As it is a flat tax, it would fully qualify for a shift to environmental taxation. Any state or provincial sales taxes of the kind would be equally suitable for the shift.

Compensation for Lower Income Earners

The increase in environmental levies would have to be compensated for in social support programs as their recipients generally do not pay tax on the financial assistance they receive and would otherwise see their real taxation increase. Lower income earners may face a similar problem because their basic deductions are large compared to their taxable income. Governments may have to look into making appropriate adjustments to the system so that their total taxation remains the same and the shift to the GEE is fair to all.

In the late 1980s, Canada went through a tax shift. The manufacturers' sales tax—a levy collected at the manufacturing level—was replaced by the GST, a retail tax. The federal government then opted to compensate lower income Canadians by issuing checks on a quarterly basis to millions of people in the country. It is still writing these to this day, some 20 years later. This is not exactly the most efficient way to do things. Adjustments need to be made directly to social support programs and the taxation system so that the benefits received and take-home pay compensate directly for the GEE.

Revenue Neutrality and Transparency

The transition to a green economic environment needs to be as voter friendly as possible. To that purpose, the cornerstone of the GEE is its revenue neutrality. For every green dollar collected, taxpayers would see their income or sales taxes reduced by one dollar. On average, the same total amount of taxes (income, sales, and environmental) would be paid by taxpayers before and after the implementation of a GEE system.

Revenue neutrality is a fundamental aspect of the GEE and is key to its political viability and implementability. Although voters may be willing to pay taxes differently, they would not be so easily convinced to give more. To ensure proper and continued political support, governments would have to be highly transparent with respect to revenue neutrality and report on it regularly.

9. Implementation Issues and Scenarios

The next few chapters will take the GEE from general principles to specifics of implementation. Concrete examples will be looked at in order to demonstrate its feasibility and dispel misconceptions. The new system will be explored from various angles and in real situations in order to get people used to the idea of being taxed differently (but not more) and show that a transition to the GEE would be much easier than it appears.

The Silent Scenario

This scenario will serve as a benchmark to assess to what extent the GEE would help achieve our goals for the environment. First, let us silence or remove the financial aspect from the green debate and see what needs to happen for the environment. Then, let us assume that we are in a perfectly planned world and that we have to draw a comprehensive plan to address environmental issues.

To conserve non-renewable resources for future generations, mineral extraction would have to be reduced. This would mean that the mining industry would decrease in size. Unfortunately, this is a necessary evil; resources cannot be conserved if we keep digging them out of the ground as fast as we currently do.

It would also mean that we would need to consume less of them. Aluminum soft drink containers would likely be phased out and replaced by glass. In many cases, tin cans for food storage would disappear from store shelves and be substituted for by more environment-friendly alternatives. Metal components in various products would be more strategically used and replaced whenever possible.

Metal-heavy sectors would see many transformations. The automobile industry would start building vehicles that are conservational, i.e. that use non-renewable resources much more sparingly. That would likely be achieved by replacing metals with substitutes wherever possible and decreasing car sizes.

Automobiles would have much longer lifespans. The used car industry would thrive as resale values would be higher and people would buy fewer new automobiles and get them fixed more extensively as a means to conserve metals. The production of new vehicles would decrease slowly in the short term as new generations of conservational, fuel-efficient, and alternative-energy cars and trucks would be needed. In the medium and long term, it would progressively grow smaller in size.

Automobiles and other vehicles would need to become more fuel efficient as well as switch to alternative energies. That has already started to happen. As time goes on, we will consume less gasoline. We will pump less oil from wells and rely more and more on biofuels and electricity.

New jobs would be created in green sectors of the economy, including renewable energy, as the demand for their products would increase. The wood and renewable resource industries would thrive and need to be properly managed to prevent their collapse as has already happened in some cases.

Feasibility

One of the main concerns with respect to economic changes is the uncertainty they can create. Scholars are still looking for the perfect theoretical solution to environmental problems, but there can be a huge gap between theory and practice, especially if the changes involved are of significant magnitude.

Brand new, maverick solutions usually face a lot of resistance. Many are never implemented simply because of the large amount of uncertainty they involve. This is a significant concern when searching for answers to environmental problems: the old ways have not worked; the new ones often present too many unanswered questions.

Although the GEE is new and sweeping in magnitude—as it needs to be, given the size of the problems at hand—it relies on an age old system. The strategy would only involve a shift between different components of government taxation. Environmental and deterrence levies are already in place in many countries. Along with income and retail sales taxes, they are all components of total taxation.

As expressed earlier, in the late 1980s, Canada had a major shift in taxation. It eliminated the manufacturers' sales tax and replaced it with one collected at the retail level. The new levy was about 7% and applied to most goods and services sold in the country. It was the first time that the latter were taxed in Canada. The transition went relatively smoothly despite being met by initial distrust and opposition. As such, we know that a taxation shift can be done.

A GEE strategy would be revenue neutral and should not change the overall amount of taxation collected in a country. For this reason, the shift would not put recessionary pressures on the economy. Consumer spending on goods and services would remain essentially the same.

For example, people in a typical country might pay $1 billion more in environmental taxes under a GEE system, but their income and retail taxes would be reduced by the same amount. As a result, their purchasing power would remain the same. In other words, things would generally be more expensive, but consumers would have more money to spend and come out even in the end. As long as the consumption level remains the same—which would be the case under the GEE—the shift should not result in a net loss of jobs.

Many environmentalists see consumption itself as *the* problem. They are correct to some extent, but there is more than one way to reduce our environmental impact: we can reduce consumption or make it greener. Opting for the former would mean a slowdown in economic growth, and few are ready to support that at the moment. Ideally, both should be done, but in the short term a greater impact can be achieved by focusing on the latter as it is much more politically viable than a reduction in consumption and the unemployment that it would imply.

The GEE would progressively make existing levels of consumption greener and greener without threatening to send economies into recession. For that reason, it is currently the best viable option we have. Under the system, growth would shift to green products. Polluting industries and the consumption of their goods would progressively change or have to shrink.

The green sector is an emerging market in which there is not a huge amount of competition at the moment. Implementing the GEE in a country would give it a significant competitive advantage in what will be a strong growth sector of the future. Export markets are likely to grow and result in a lot of job creation.

The GEE is essentially based on established and well-understood economic practices. All countries around the world have lived with the ins and outs of taxation for a long time. The strategy is readily implementable and has a relatively low level of uncertainty, given the magnitude of the changes that would result from its implementation.

Geopolitical Concerns

The ups and downs in the price of oil over the last few decades have given us some insights on the geopolitical implications the depletion of resources can have for modern economies. The conditions created by shrinking supplies increase poverty and the cost of living and can trigger recessions as well as deadly conflicts.

A global move to resource conservation could significantly extend reserves and slow down the depletion process, easing up the pressure for potentially chaotic events in the future. Mineral resources other than oil have been less a concern to most countries because their supply has remained largely uncartelized and unpoliticized. They are also less massively used compared to energy.

However, metals are much less replaceable than oil. Alternatives to petroleum do exist and are actually plentiful. They are just more expensive. That is not the case for other mineral resources. What will happen when their supplies begin shrinking like petroleum? They are a major component of the infrastructure of most countries, and our lifestyle depends on them.

At the current rate of depletion, shortages will soon begin to occur, and the economics of metals could resemble those of oil.

Power could shift to mineral-rich countries just as it did for the Middle East with respect to energy. Countries that import much of their mineral resources will see power shift away from them, their wealth quickly following suit.

That would likely be the case for the U.S. The question is, do we want to accelerate this process with high rates of depletion and precipitate economic decline and more global crises, or do we want to conserve resources worldwide and decrease the potential for problems developing as a result of scarcity?

There are only two choices available to us. We can do nothing and let the process occur haphazardly and end up with a repeat of the Middle East power shift and related human suffering, or we can make it happen within a global framework that would provide for sanity, prevent cartelization, and ensure continued supply to all countries.

The long-term self-interest of all countries lies in starting the conservation of resources immediately and in supporting international efforts in the same direction. Europe and the U.S. will be especially vulnerable to future crises because their economies and predominant lifestyles result in the use and waste of a lot of resources and they have already exhausted much of their own local supplies.

Countries will want to cut back production and use of metals by a fair amount in the medium term. Significant changes are needed. The reality of non-renewable resources is one of impending crises that can be softened and delayed, perhaps avoided, with the proper global framework.

Rate of Implementation

Two of the issues regarding the conservation of non-renewable resources are how far and at what rate we should cut back their extraction. No single person can determine that. In the long term, we would have to be looking at extended sustainability levels. In the immediate future, the answer will likely come as a result of a political and social process. A couple of considerations will be useful in helping with the decision process.

The potential job losses resulting from the implementation of a new tax structure is a legitimate social issue. Large-scale unemployment is not expected to result from this as the total level of taxation would not increase and countries' money supply would not change. Overall, the same amount of consumption and investment would be around, but it would shift from one sector to another, for example, from oil to renewable energy.

Total employment should therefore remain relatively stable, but work would not transfer directly from one sector to another. If you lose your job as an oil worker, you may not get one in the new wind turbine industry or get in at the same pay rate as you had before. So, just how fast should we go?

Attrition has been used in the past as a worker-friendly approach to change and is usually one of the better alternatives whenever possible. The way it works is that rather than laying off people, companies do not hire new employees as older ones retire or quit for one reason or another. Alternatively, they offer bonuses to workers willing to take early retirement. These are undeniably the best possible options when companies can afford them.

However, neither system is perfect. When a business closes down, there is no money for early retirement and everybody gets laid off, old and young. Even when planned, a closure by attrition or early retirement leaves casualties.

How long would it take to fully implement the GEE if we did it by attrition? New entrants into the workforce may average about 20 years of age. Let's assume that most will retire at age 60. That leaves an average career span of about 40 years. Add another 10 years as a measure of security. As such, someone starting work today would technically retire in 50 years.

Consequently, at the best possible speed, we would need about 50 years to bring down production amounts of non-renewable resources from 100% (current levels) to 0% if we were to totally stop the extraction of minerals. Obviously, that is not what we are trying to do. If we were to bring them down to 50% of current levels in the medium term, it would take us about 25 years through attrition.

The above is not to suggest that we should reach that specific target in that span of time, but it is a benchmark that can help us see more clearly into the future and minimize potential socio-economic disruptions from a GEE implementation. We may choose to go faster for environmental expediency and especially at the begin-ning, or more slowly. Ultimately, the socio-political process will determine the exact speed of implementation of the GEE.

If half of one attrition span (25 years) is deemed both a reasonable and responsible time line to reach medium-term resource conservation targets, by 2035 we will have cut by 50% our consumption of non-renewable resources. Another half-span would take us to 25% of current consumption levels by 2060.

What targets countries will chose for the very long term is uncertain at this point. What we know for sure is that we should start early. Our current rate of use is highly harmful to the environment and severely depletes resources for future generations. The sooner we start, the better.

Taxation Management Concerns and Solutions

Environmental initiatives can pose a number of problems if not designed appropriately. Alain Verbeke (Solvay Business School, Free University Brussels, Belgium) and Chris Coeck (Faculty of Applied Economics, University Centre of Antwerp, Belgium) expressed concerns that poorly managed environmental levies may result in a backlash in the business community, a decrease of their support for environmental taxation and the general impression that environmental taxes have become "arbitrary measures to stabilize government income" (Verbeke and Coeck, 1997, p. 510). In other words, politicians could abuse the system to raise total taxation.

The GEE is specifically designed to be revenue neutral. The proposed system is a shift and not additional taxation. As such, it is meant to not be used to arbitrarily increase government income. Revenue neutrality is a large part of its political viability. Open disclosure and a high level of transparency would be part and parcel of the new system and would easily prevent abuse.

Concerns have also been raised that green taxation is often used to fund environmental programs (for example, taxing contaminants and using the proceeds to fund water improvement programs) and that the environment would get disproportionate amounts of funding while other programs would come up short.

This is not the case with the Green Economic Environment. All new levies would be directed to general revenue as is the case with current income and retail taxes. This would prevent the arbitrary financing of certain programs over others and ensure that government spending on social programs and services remains the same.

The GEE is not about increasing funding for the environment at the expense of other programs. It is about creating a green incentive structure for economies.

Another concern raised in literature with respect to taxation is its dynamism. Verbeke and Coeck (1997) warn that using taxation as a source of income for governments or as funding for environmental programs may not yield the intended benefits for a number of reasons. Green taxation revenues tend to be dynamic and may not provide as stable a source of government income as desired. Let us look at an example of this.

Suppose that a government implements a carbon tax that is expected to generate a billion dollars in revenue annually. The resulting increase in gasoline prices would generate positive environmental behaviors as expected—the purchase of more energy-efficient cars, increased use of public transportation, a switch to alternative energies, etc.

Because this would result in a decrease in oil consumption, over time the tax would bring in less than the expected revenue target. Governments would then have to increase it further to generate the originally desired one billion dollars. This action would presumably lead to lower consumption yet. The environmental tax would have to be raised again. Dynamism could result in a kind of *treadmill effect* which would eventually create a number of problems. Let's address these issues.

The problem that Verbeke and Coeck describe is a very legitimate concern. Continuously increasing levels of environmental taxation could seriously undermine the strategy as well as a country's international competitiveness. However, the dynamism of environmental levies is a very good thing in itself. It would spur us on to continue year after year to improve on resource conservation and environmental protection. We want the economy to become more and more environmental over time. Dynamism is actually the chisel which would make society and the planet greener and greener year after year. But it needs to be controlled.

The implementation of the GEE would be progressive. This implies that initial taxation levels would be lower and need to increase in order to reach long-term target levels. As such, there would be room for raising tax rates without creating problems, and dynamism would therefore be an integral part of the implementation process.

Once desirable levels of conservation and protection are achieved, or international competitiveness ceilings have been reached, we can reverse the process and raise income tax if the government take from environmental levies falls. Of course, that would have to be done without increasing total taxation. As such, the dynamic effect of the GEE is perfectly and easily controllable. We just have to remain aware that the issue will need to be managed.

Note that under the current system, government revenue varies from year to year depending on economic cycles, profits, personal earnings, unemployment rates, national retail sales, etc. It decreases in times of recession and increases in periods of growth.

Varying revenue is not a new issue for governments; gaps in income have historically been made up by temporarily borrowing money or through budget cuts. The GEE would only continue that, not change it.

National and International Issues

International competitiveness is probably the most difficult issue concerning the implementation of a green economic environment. Although most countries could readily initiate and gradually put in place such a system, few would be able to quickly proceed with a full-scale implementation because of the effect on their ability to compete internationally.

The GEE is not unique in this respect. Minimum wage levels, interest rates, productivity, social programs, greenhouse gas reduction targets, and a number of other factors also affect international competitiveness. The GEE would be just one of a range of variables in that respect.

Any resource tax would by necessity be collected prior to export and make a country's goods more expensive and less competitive in foreign markets. Although exceptions could be made, a rebate on exports would not be a realistic solution as a resource tax would become diffused in finished products.

For example, while a bar of steel may see a full increase in price from a given levy, a product with a 30% content of the metal would see only a partial rise in price. For that reason, it would be far too complex to try to estimate the percentage of non-renewable resources in every item to be exported and compensate for the levy with a rebate.

Individual countries would therefore be limited in their ability to tax resources past a certain level as it would overly decrease their international competitiveness. As such, most should be able to carry out the initial phase of the GEE as it involves lower tax rates, but trade-bloc or international accords would likely be needed to support a high level of implementation of the system.

That being said, even a milder variant of the Green Economic Environment would sill yield substantial benefits. There are obviously the positive effects with respect to carbon emissions, contaminants, non-renewable resources, the green energy sector, etc. However, there are other advantages to the system as well.

For example, it would mean that fewer non-renewable resources would be exported to be wasted overseas, therefore preserved for future generations. Countries would not want to conserve and tax resources in home markets but export them cheaply to places where there are no conservation efforts being made.

As well, the GEE would affect primary resources more strongly than final goods, as seen in the *bar of steel* example above. This would prompt countries to transform them more extensively into finished products and develop their local manufacturing industry—which would create jobs—rather than export them as raw materials.

Keep in mind that countries could choose to not fully tax the export of raw materials in order to lower the impact of the GEE on their international market shares. To soften the effect of resource levies on certain industries, governments would also have the option of initially imposing import taxes on big ticket items so as to maintain local manufacturers' competitive advantage within a country. As more and more countries get involved in conservation programs, import taxes could be progressively phased out.

Note that international competitiveness is an issue with greenhouse gases as well. The establishment of common carbon emission targets (within groups of countries) in the Kyoto Accord addresses the problem. The same solution could simply be applied to the GEE.

The Green Economic Environment strategy could work with already established emission targets and, as stated earlier, alongside cap-and-trade. It would simply act as a mechanism (which neither the latter nor Kyoto provides) to reduce greenhouse gases and enable countries to meet their carbon reduction commitments.

An emission trading system is only a flexibility component which allows companies that exceed the caps agreed under Kyoto to buy credits from businesses that have done better than their reduction limits. You do the math! Plus one thousand units minus one thousand units equals zero. As a whole, no net carbon reduction is achieved by the system. Incentives and decreases in emissions are the result of the Kyoto Accord's caps.

When you look at it closely, most countries do not have a mechanism that would foster or help the industry and consumers to reduce carbon emissions. The GEE does provide one. As discussed later on, the Green Economic Environment could replace both the Kyoto Accord and cap-and-trade and deliver a simpler, less costly but much more powerful market-based approach for carbon emission reduction whether at the local or the international level in the form of a new global environmental accord.

The long-term and most efficient solution to the problem of competitiveness differentials is the negotiation of international accords. Under such agreements, countries would see their taxation levels increase at the same agreed upon rate just as they do with carbon emission levels under the Kyoto Accord.

Because of its scalability, the GEE would be readily implementable on a national basis in most countries around the world. The first ones to implement the system would also be the first to shift their economies towards cutting-edge sectors of the future (metal substitutes, conservational techniques and designs, alternative energies, electric and fuel-cell vehicles, etc.) and would therefore gain a huge competitive advantage in wide-open emerging markets.

International agreements would remedy to competitiveness problems between countries and make it possible to shift into high gear the fight against global warming and the worldwide conservation of resources and preservation of the environment.

Burden of Change

Throughout history, all too often the cost of social advancement, the burden of change, has been borne by those who lose their means of living as a result of economic events or industrial and social shifts. To a large extent, this is a choice. If society as a whole benefits from certain changes, why should a few be responsible for the costs associated with them? Why should we not compensate them for these?

Clearly, some of the costs of change cannot be avoided, but we, at least, can mitigate them. It is not only an issue of social justice; the absence of compensation is often counter-productive. The higher the unmitigated costs, the more reluctance to change is created in a society.

In a constant race for economic positioning, a nation which resists change loses its edge and quickly falls behind. Tomorrow's innovative societies are those which will be best able to quickly adapt to new and changing environments. Removing blocks to change is becoming an increasingly important factor in this rapidly evolving world.

Unemployment in particular sectors or for specific people is often a direct cost of change. Improving the way we deal with this problem would ensure that certain individuals do not unduly bear the burden for changes that benefit an entire country. In fact, it is the responsibility of a fair society to make certain that they do not. It would also help remove resistance to change, making a country much more innovative and successful in a competitive world environment.

Most countries already have unemployment insurance programs or other types of support systems along that line. If we are serious about environmental change—we know that the needs are massive —these social safety nets have to be given a second look so that they act more in concert with the changes that need to happen in society and target more closely the sectors that may be affected by new policies.

Better support would make the implementation of a green economic environment easier and benefit countries in the long run.

10. A Blueprint for National GEE Implementation

This chapter looks at the steps for the implementation of the Green Economic Environment strategy at the national level. But first, it runs through a theoretical scenario to explore the ins and outs of the system.

Implementation Scenarios

The next few sections take a closer look at theoretical scenarios of implementation for the GEE. This will help visualize the new landscape and demystify certain issues. The first one is a sudden and drastic script that will serve to dispel some of the myths that may arise as a result of the work of lobbies opposing environmental change. The purpose of this extreme scenario is to enable the testing of the strategy's feasibility. Of course, it is not meant to be a viable option for countries but rather to stretch ideas to the limit.

Canada is used as an example as the country's size makes it easier to illustrate situations. Monetary figures may be viewed in either U.S. or Canadian dollars as both are about equal in value at this point. As well, for the time being, issues of international competition are set aside so as to be able to focus strictly on a national scenario of implementation.

D-Day Implementation

D-day implementation refers to a drastic scenario which involves high initial tax levels. It would occur on a national basis and, as already expressed, assumes no international trade issues.

Suppose that the Canadian government decides to implement a Green Economic Environment strategy overnight. The federal retail sales tax on goods and services (GST) as well as its provincial equivalents would be eliminated. This would automatically lower the price of goods and services by about 12%.

The non-progressive bracket of income tax (its flat component) would be dropped. As such, the first $30,000 of net personal revenue would no longer be taxable, and the average worker in Canada would save about $10,000. As a result, that person's monthly take-home pay would be instantly about $800 more ($10,000 divided by 12). Someone having a gross revenue of about $40,000 would have about $30,000 net after the basic exemption (about $10,000) and, therefore, would not pay any income tax.

The new GEE levies implemented by the government would include: a 100% tax on non-renewable raw materials (effectively doubling the price of steel, copper, nickel, etc.) to promote conservation; a flexible tax added to the cost of oil to maintain its price around US$ 150 a barrel (or any other price determined appropriate, politically or otherwise—which could be more or less) in order to reduce gasoline consumption and greenhouse gas emissions and promote the development of the renewable energy sector; and a 100% tax on key contaminants and pollutants.

Supported by regulations, standardization would be used in the packaging industry to further encourage resource conservation. Reusable containers would be tax exempt. Those that cannot be reused but are recyclable or made of renewable materials would see a $1.00 levy. Those that are not reusable, not recyclable, and not made of renewable materials would have a $2.00 packaging surcharge added to their price.

After a few months, the resource tax on metals would have filtered up. The impact of the GEE on particular products would depend on a number of factors: the content of metals, the amount of transformation that raw materials go through, the proportion of labor and technology in the product, and the total final value of the item. On average, consumer products with high metal contents would have seen a 25% to 35% increase in price while the costs of those with moderate amounts would have risen by 15% to 25%. Items with low metal contents would have seen an increase of 5% to 15%.

Computers and consumer electronics—which can have 80% to 95% of their value coming from the cost of technology (the operating system, memory, the CPU, etc.)—would not have gone that much higher in price although they can contain a fair amount of metal. Luxury cars would have been less affected by the new levies than regular ones as a larger proportion of their costs comes from labor, which reflects itself in a much higher final retail price.

Household cleaning and other chemical-based products that are harmful to the environment would have almost doubled in price because of the contaminant tax. At the end of the year, the taxpayer who had been $10,000 richer because of the income tax reduction would have spent $10,000 more on purchases. The government would have given from one hand and taken back with the other!

The purpose of the GEE is not to raise more money for the environment or the government. It is to change the incentive structure of the economy so as to naturally deter unenvironmental practices and promote green consumer behaviors. On average, people would break even at the end of the year, but consumption patterns would change for the better. As unenvironmental goods become more expensive, people would move away from them and shift to greener alternatives.

The GEE would create a green economic environment in which one would be rewarded for doing the right thing (buying green) and punished for purchasing products that are not environmental. And all of this, at no additional costs to the taxpayer.

The marketplace would also significantly change for businesses. Unenvironmental products would become hard to promote and less and less profitable. Green goods, which had been difficult to sell because of their higher prices, would gain a competitive edge and see their markets expand. Many products would be redesigned with less or no metal as the manufacturing input became more costly. The demand for automobiles would shift to smaller fuel-efficient and conservational sizes and cheaper models that used substitutes for metals whenever possible. Green industry sectors would be on their way to becoming highly profitable.

A drastic overnight implementation of the GEE would be difficult but manageable. Some things would become more expensive, but people's disposable income would increase in proportion as a result of people having to pay less income and retail taxes. As such, total consumption (hence, jobs) and *purchasing power* (the total amount of goods that one can afford to buy) would remain the same. People would be spending differently, not less. Unemployment would not have risen. The way of life would change, but the overall standard of living would be comparable to earlier on.

Under the GEE, we would see the world around us become greener and greener, year after year. Less metal would be part of our lives. Contaminants and toxic compounds in products would decrease. Fossil fuel use would go down, less garbage would be produced, the environment would begin cleaning itself up, etc. We would see an explosion of alternative energy technologies appear and a battery of green products hit store shelves, most of them cheaper than their alternatives.

Habit Shifts

Some people would likely want to take maximum advantage of the new green economic environment provided by the GEE and radically change some of their habits. For example, they would consume a lot more green products and keep their cars longer or get much smaller and more fuel-efficient ones. Their new environmental habits would lead to significant savings. In the example above, their actual expenses may only increase by $7,000, leaving them with a $3,000 bonus from the $10,000 tax break they received on income.

Others would go with the flow and change their habits at about the same rate as most people. They would spend $10,000 more, the same amount they got back in income tax reduction, and come out even. However, their habit changes would be for the better. A lot of resources would be conserved, carbon emissions would go down, and contamination would decrease.

Some people simply do not care about the environment or can afford not to. They would continue to change cars as often as before, fill up landfill sites, and buy products in non-renewable packaging. They would find life more expensive, spending more than before on consumer goods.

Overall, the GEE would be neutral, but some would benefit more than others. The difference is that under a green economic environment, those who do the right thing for others and future generations are the ones who would win out.

Employment Shifts

Assuming no international competitiveness issues, an implementation of the GEE would not cause widespread job losses as total spending would not change. Money would not be taken out of the economy.

Unenvironmental sectors would see employment decrease, but green industries would have a corresponding increase. The money not used to purchase goods harmful to the environment would now be spent in other sectors of the economy. Job creation in these would generally make up for the losses from non-green industries primarily because the GEE is revenue neutral.

Overall, because the amount of consumption has remained the same, there should not be any increase in unemployment as a result of the green economic environment strategy proposed in this book.

The Forest Industry

Just as some sectors in the economy would decrease in size and importance under a GEE system, others, such as the forest industry, would thrive. The demand for its products would increase. That would mean growth and more jobs. Government regulations would need to be put in place to ensure the proper management of the industry.

For example, clear-cutting practices would have to be phased out and replaced by better options. Replanting would have to increase to not only renew the resource but also expand it to meet increased demand. Forest preservation for current and future generations would have to become a priority. Regulation will likely remain the primary instrument for the protection of old-growth forests.

What Needs to Happen

Let's look back at the initial planning exercise that we did to determine what an optimal environmental plan would look like (the silent scenario).

The GEE would result in higher costs for products with high metal contents. Manufacturers would shift to substitutes whenever possible, and the automobile industry would start building more conservational and environmental vehicles. All of this would help save non-renewable resources, which is what would take place in an ideal situation.

Products that are made with toxic chemicals or are harmful to the environment would become more expensive under the GEE. Manufacturing and consumption would move to greener alternatives. Carbon taxes would result in a shift to renewable energy. As a result, contaminants and carbon emissions would decrease, which are things that would also happen under an optimal environmental plan.

The GEE strategy proposed in this book would deliver exactly what would happen under ideal circumstances. Of course, some job and economic displacement would be inevitable no matter how we decide to address environmental problems. A better unemployment insurance system would help us make the transition and result in a society being more competitive in the long term.

The GEE would make the changes much less painful than the other options we have. The D-Day scenario was one of drastic implementation, but countries will likely begin slowly, making the transition relatively easy.

We have the means to bring about large-scale environmental change; the only question remaining is, do we have the will? We do not really have a choice. Problems are already bad and will only get worse with further delays. The GEE will inevitably result in a number of lifestyle adjustments. However, together they will have a huge impact on global warming, conservation, and the environment.

We can act now; the GEE gives us the means to do so.

First Step: Laying Out the Foundation

The GEE is scalable. This means that the initial tax rate on raw mineral resources could be 100%—a doubling of current prices—or 30% or less as necessary. As such, the Green Economic Environment strategy could easily be phased in slowly and progressively, making it implementable without an excessive amount of planning. In the beginning, it would likely be difficult to estimate the most appropriate level for each tax. However, because the system is scalable, it would not be necessary to do so.

Lower initial rates would decrease the impact of the GEE on the economy and allow governments to see how environmental taxes interact with each other and to assess appropriate levels. They would also help familiarize the public with the implications of the new system. As such, scalability would make it possible for countries to begin creating the green economic environment without delay. Of course, as environmental taxes are collected, the retail sales levies and the non-progressive share of income tax would be reduced in equal amounts.

What is important is to lay down the foundation as early as possible to start reducing carbon emissions, the depletion of non-renewable resources, and contaminants. Initiating the implementation early would have several advantages: it would give us more time, allowing for a softer implementation, and dispel any hopes that a government commitment to a greener future is lip service. It would be a firm signal to the business sector that times are definitely changing and that companies should start planning for a green economic environment.

Second Step: Short-Term Levels of Taxation

The second stage is intermediate in nature and would seek to establish a realistic national implementation. Countries would more fully commit themselves to the new system and select rates of taxation should be high enough to achieve significant amounts of resource conservation and pollution reduction but not so high as to overly affect a country's international competitiveness.

Third Step: Progressive Implementation

This step would lead to the establishment of optimal national levels of taxation for the different categories of products and resources involved. It would essentially be a fine tuning of the tax rates developed during the second step. The resource and other GEE levies would be gradually increased to the new targets.

Fourth Step: International Agreements

As national implementation would likely reach limits, the fourth step would be the development of international agreements. These would support fuller national implementation, eliminate international competitiveness issues, and take the green economic environment worldwide.

The GEE Diffusion Effect: A Closer Look

The GEE would be charged on raw materials and other products and diffuse itself as it moves up into finished products. Even fairly high initial tax levels—if this is what a country chooses to do— would not result in excessive price hikes for consumer goods. This section takes a closer look at the issue.

Tax Diffusion Through the Economy

The GEE would spread in different patterns throughout a country's economy. At one end, items like machinery, machine tools, and automobiles would probably be hardest hit by resource taxation as they contain large amounts of metal. However, a drastic 100% tax on raw materials would not automatically mean a doubling of the price of these items. The parts and components of machinery and vehicles are not simply metal. They are technology, labor, and materials.

A kilogram of steel and a tractor part weighing one kilo are not the same thing. The part is metal that has been smelted, cut, or shaped into a specific design. In those processes, value is added to raw materials. For example, the part weighing a kilo may sell at double the price of its weight in steel. That tax would therefore apply to only half of the final price. Let us look at an example based on the drastic implementation scenario discussed earlier.

Suppose that a tractor made of 90% metal had a value of $100,000, with the cost of its metal parts being $90,000 and that of its non-metallic components, $10,000. Firstly, the tractor would not double in price from a 100% tax on raw materials as only its metal parts would be affected by the tax; the $10,000 worth of non-metallic components would not go up in price.

Secondly, the tax would affect only the raw materials, not their fabrication. Suppose that raw materials cost $40,000 and their manufacturing into parts, $50,000 (for a total of $90,000 as above). A 100% tax on non-renewable resources would double the cost of metals from $40,000 to $80,000. The costs of manufacturing the parts ($50,000) and of non-metallic components ($10,000) would themselves not be affected. The final cost of the tractor would have increased by $40,000 to $140,000, which is a 40% rise in price.

At the other end of the spectrum would be services, such as the legal, educational, and healthcare industries. As these do not generally involve the selling of material goods, their costs should essentially not go up. If anything, they should decrease as retail taxes are dropped. The same would be true for renewable resources such as wood products and foodstuff.

In between, you have the vast range of consumer products that would see varying increases in price depending on the value of the non-renewable raw materials they contain as well as the amount of transformation these have gone through.

A New Consumer Environment

Going back to the Canadian example, dropping the federal retail sales tax would make goods and services 5% cheaper. Doing the same thing with the provincial sales taxes would further reduce most prices by another 6% or 7% for a total of about 11% to 12%. As a result, under a 100% GEE scenario you may see the price for high-metal content goods increase by about 25% net. The cost of services and renewable resources would decrease by a few percentage points. Other items would range from no change in price to moderate increases.

Many things would be cheaper, others, more expensive, but our total purchasing power would essentially remain the same as we would have more disposable cash from the rebate on income and retail taxes. The GEE would mean a different way of life but not a lower standard of living. We would experience a different consumer environment which would lead us to buy fewer unenvironmental goods and more green ones. Our buying patterns would change.

The Packaging Scenario

In the packaging industry, the strategy would yield similar results. The GEE would, for example, bring in taxes on new containers. The environmental levy would optionally be backed up by regulations standardizing them. The products purchased by consumers would remain exactly the same. A fruit juice is the same regardless of what it is bottled in.

As all new containers would be taxed, companies would naturally and progressively shift to recyclable and reusable alternatives. The soft drink and beer industries in many countries used to function on that basis and still partly do so today when they reuse their own bottles. In terms of economic organization, this is essentially old technology.

What would be different under the GEE is how the recycling of empty containers would work. It would be based on markets as opposed to the bureaucratic refundable-deposit system. Empty bottles would eventually be *sold* to recyclers rather than *returned*. Of course, the old way of doing things would still remain an option in situations where it is desired or is still the best alternative. Under the GEE, choice would still exist for consumers. Soft drinks in plastic bottles would still be available but more expensive because of their less environment-friendly packaging.

One thing that would change in our lifestyles is that we would likely spend more time recycling, taking items over to depots or processors and selling them back for reuse or raw materials. Those who would be too busy to do so, or would not want to, could forgo the cash and just leave them at the curbside to be picked up by recycling entrepreneurs or kids wanting to make a little cash.

The Dynamism Issue

The chisel effect of the GEE in shaping a new society would be continuous. Dynamism is a very positive and desirable factor in bringing about environmental change. Metal content would decrease in many things as non-renewable resources would be substituted for by renewable or reusable materials. Businesses would continue to seek ways to lower their costs by switching away from taxed inputs. As the price of toxic intermediate chemicals would go up, they would be increasingly replaced by more environment-friendly alternatives. Processing methods would change and become greener.

A continuous incentive to do better is exactly what we want. For the first time with respect to the environment, the issue would not be the lack of funding but keeping positive change from happening too fast.

11. Disposable Grandchildren: Packaging and Contaminants

This chapter will cover in more details one of the most important sectors of the economy as it relates to the environment: packaging. Its products are pervasive in society. Furthermore, many are single use and, for that reason, extremely wasteful. As such, the packaging sector has a massive impact on the environment and is in need of major changes.

Currently, conservation includes various recycling and reuse programs. Renewable resources such as paper products are also the focus of recycling because of the cost of their disposal. The packaging industry is of direct relevance to resource conservation not only because of its use of non-renewable materials—such as steel and aluminum—but also because of its products' utilization of landfill space. Targeting the sector is crucial because of the sheer amount of waste it generates and because, if properly managed, it is one of the keys to resource conservation.

The 20th Century Approach

How should we manage the packaging industry? Should we impose severe restrictions on it? Should we tax it or assess import tariffs? It is undeniably one of the most difficult environmental problems to handle. The way we currently manage packaging is nothing short of a crime against humanity.

Packaging serves once and is then discarded. How can we still be so widely using non-renewable resources for its fabrication? We consume tons of depletable materials—which will be desperately needed by future generations—for products that not only see virtually no use but also cost a lot of money to dispose of and will plague landfill sites for a long time!

The 20th century can make one claim: to have brought about the *disposable society.* By indulging in convenience, we are turning the world into a garbage dump, making our very grandchildren disposable. They will be left living in a highly polluted environment, with their own body tissues reflecting the chemical mix around them.

Recycling and reuse programs will need to see significant shifts in approach and scale if they are to achieve effectiveness and produce reasonable results for the environment. Recycling is not the long-term solution to resource conservation. It is just a part of it. On the current scale, it only mitigates the problem although we may feel that a lot is being achieved.

Plastic soft drink bottles, for example, can be recycled to make t-shirts, carpeting, or pillow stuffing. However, most bottles do not get recycled in the first place and end up in landfill sites. Products made from recyclables eventually also end up in garbage dumps, only later. Steel containers can be resmelted, but again, a lot of them do not get recycled in the first place, and even metal that has been given a second life eventually rusts away into the environment.

Recycling helps conservation, and efforts in this direction should continue. However, it only slows down depletion to some extent and delays the inevitable. For that reason, the real long-term solution for the industry is to reduce, reuse, and shift to renewable resources such as paper and cardboard.

The Market Approach

Recycling programs often have to be funded by governments because there is generally no market for used materials at the actual cost of collection. That is, recyclables are most often resold by municipalities at less than it costs to pick them up. Taxpayers, therefore, have to make up the difference. Furthermore, the government bureaucracies that run the programs tend to be less efficient than their private sector counterparts.

Another approach to recycling has been the refundable-deposit system in which a small levy is charged on bottles and cans and refunded when the empty containers are returned to vendors. Hundreds of millions of deposits have to be collected and kept track of by retailers. Then, each has to be refunded. Net surpluses and deficits have to be accounted for and returned to or claimed from government agencies. Again, this may not be a very efficient approach.

The GEE would enable and support a new strategy for conservation. A levy on packaging and/or materials would have the double effect of reducing our consumption of non-renewable resources and of creating natural markets for reusables and recyclables by increasing their value above the cost of their collection, i.e. by making the industry profitable. This could replace the more bureaucratic and inefficient deposit system although the latter would still be an option wherever it is more functional.

Under the GEE, private companies would buy and sell reusables and recyclables for profit. Items would be collected and sold for cash at market prices without government intervention. The system would have the advantage of keeping the element of choice for both corporations and consumers. Certain useful but wasteful types of packaging would still be available, although for a higher price, as opposed to being regulated out. This would provide more options for consumers and is generally preferred by businesses as it gives them more flexibility and time to adapt.

In the long term, natural markets would develop by themselves as non-renewable resource taxes are progressively increased and profitability levels are reached in those sectors, i.e. when the cost of recycled resources such as metals is significantly lower than that of newly extracted minerals. However, in the short term, there could be the need for a combination of approaches.

The Short-Term Market Approach

Although resource taxation remains, in my opinion, the most efficient approach to environmental change, its initial levels would be limited by international competitiveness. As such, an interim strategy for packaging is likely to be needed in the beginning.

The goal for the industry would be to shift to either reusable containers or 100% renewable, recyclable, and biodegradable resources such as cardboard. Non-renewable and non-biodegradable materials such as polystyrene fillers (better known under the trade name Styrofoam) would be taxed in order to foster their replacement by environmental alternatives such as cardboard frames, molded paper pulp, etc. Regulations could further be applied to inks, glues, tapes, chemicals, etc. to ensure that they are of only non-toxic and fully biodegradable types.

With respect to food-grade and other containers having a potential for reuse, the GEE would tax new items, making used ones or those made with recycled materials cheaper. Market forces would naturally act to shift industries towards them. Food-grade containers include the various plastic tubs and steel cans that edibles come in as well as the aluminum cans and glass or plastic bottles used for liquids. The goal would be to switch from them to reusable alternatives as they are a large part of our daily production of garbage.

Generalities

Under the GEE, people would not return their empty containers for a refund. Instead, they would resell them to a local recycler for what these would be worth on the market at that time.

The role of governments would be to set taxes sufficiently high so that empty containers would be worth reselling instead of being thrown away. In some ways, the GEE approach would resemble the refundable-deposit system, except that it would not have its bureaucracy. One of the main differences between the two would be that the price of returns would not be fixed but vary from recycler to recycler and by locality.

Communities that are too small for recycling to be profitable or that do not have local recyclers could continue the programs that they already have.

Double-Taxing Inputs

Under the general GEE scheme, raw materials (outputs) from producers would be taxed once. One way to solve the lack of initial incentive or profitability in the reuse and recycling industry would be to tax those outputs again when they are bought by packaging manufacturers (as inputs).

For example, steel and glass would be taxed once with producers at the established GEE rate. Manufacturers that use these to make regular goods (tools, toys, dinnerware, etc.) would not pay a second levy. However, companies using them for the manufacturing of packaging would be taxed a second time as the raw materials are purchased as inputs.

This would make containers fabricated from new materials more expensive, which would increase the incentive for packaging manufacturers to move to used materials. It would also create a market for recyclables since waste metals would be cheaper as they would not be taxed. Manufacturers of packaging would naturally shift to them.

Each type of material could be assessed for environmental friendliness or desirability. Criteria such as reusability, renewability, biodegradability, and toxicity could be used to set tax levels. For example, steel and aluminum would be assessed higher levies because they are depletable. Cardboard would be at the bottom of the scale because it is both renewable and biodegradable. As desired, the various types of plastics currently used in the packaging industry could be assessed individually and get different tax rates.

Imported containers could easily be taxed based on weight. Governments would simply have to require exporters to list the types of materials and net weights of packaging on labels and shipping documentation.

This would probably be the simplest approach to creating a market-based strategy for the packaging industry at the beginning. In the longer term, the basic GEE output tax might be enough to support profitable markets for used materials although input taxation would remain an option to ensure high standards of recycling in an industry that is very wasteful.

Keep in mind that we are still discussing a revenue-neutral approach in which higher environmental levies would be offset by a drop in other taxes.

Individual Taxation

A less desirable strategy would involve individual levies on types of items. This approach would be more difficult to implement because of the variety of packaging available. However, it may be preferred by some countries for one reason or another. Let us first look at the range of available options.

One of the best environmental packaging at the moment is glass. It is natural, recyclable, and reusable. Under an individual taxation scheme, new containers would face levies to promote reuse and expand markets for recyclables. Used ones would remain unlevied.

Plain cardboard would be a very good choice too but can only be used for packaging dry goods. For liquids, an alternative to glass is waxed cardboard cartons as are currently used in the packaging of milk and some drinks in many countries. These, however, are generally not reusable. They could be levied but at a low level because of their bio-biodegradability and the renewability of their materials.

Other alternatives such as regular plastic bags (those used for carrying groceries) and plastic-lined cardboard cartons could be levied at an intermediate level. They are less environmentally desirable but better than many other options.

Because they do not bio-degrade easily and are generally not made from renewable resources, plastic containers would be fully targeted under an individual taxation system. So would steel and aluminum cans. However, the same items made from recycled materials would be taxed at a lower level in order to promote their use and the development of their markets.

Examples of Individual Taxation

Individual taxation would be fairly complex. The following is an example of it and assumes that regulations standardizing container types and sizes have been implemented (see *Standardization* a little further down).

For the sake of simplicity, the following will deal only with smaller size items such as soft drink bottles and steel cans for foods. Larger or more expensive containers could be the object of a separate scheme or category. *Non-renewable* would refer to packaging made from resources that are limited in supply and depletable such as steel, tin, aluminum, etc. As most materials do eventually biodegrade, *non-biodegradable* would refer to those that do not break down within a few years in nature.

A typical scheme could look like this. Taxes would be charged preferably directly to manufacturers and on imports at customs. The first level of levies, $1.00/item for small sizes, would be applied to new non-renewable, non-reusable, non-biodegradable containers.

It would essentially comprise items that we would want to phase down or out for their unenvironmental or unconservational characteristics. This would shift producers and food processors—as well as consumers—away from them and towards better alternatives. The $1.00 levy would not be a refundable deposit but a cost. It would be recouped through revenue-neutrality and by selling containers back to recyclers.

The second level of levies, $0.75/item, would be applied to composite containers such as plastic- or foil-lined cardboard cartons, for example, those currently used for packaging juices. This level would represent better alternatives. The levy would promote a switch to more environmentally desirable types of materials.

The third level of taxes, $0.40/item, would be charged on all new non-standard containers not already taxed above and made of reusable and recyclable materials. This would represent good alternatives such as glass containers. New and non-standard items would be taxed to promote a shift to standard ones, which would be more reusable, and to encourage reuse.

The fourth level would target new standard containers. A $0.30/item levy would promote their use over that of non-standard ones. The less fragmented markets are, the cheaper and more efficient processing for reuse would be because of economies of scale. This would result in lower costs to consumers, more successful processing industries, and increased resource conservation.

Fifth level packaging, used containers, would not be taxed. At present, this field is fairly limited. Beer is one industry in which bottles are washed, sterilized, and reused. However, this is undeniably the exception rather than the rule. Various foods could be packaged in reusable glass jars. Soft drinks could go back to being bottled as they used to.

The above could be a typical packaging scheme used to promote environmental and conservational behaviors in the sector. Of course, rates could be higher or lower. As a general rule, the higher a tax, the more industries would shift away from undesirable products and towards more environmental alternatives.

Countries could distinguish between types of plastics based on reprocessability. Reusability could also be graded and levied differently based on the number of times a container can be refilled. Ultimately, each container type could even be graded individually based on environmental and conservational characteristics and desirability. So could raw materials under a double-taxation scheme.

Overall, the above system would mean that the choices that are better for the planet would be less expensive. Plastic soft drink bottles would still show up on shelves; so would steel and aluminum cans. However, they would be more costly. That would provide for flexibility for both consumers—who would still be able to choose lighter packaging out of convenience—and producers, which may find it cheaper or more useful to use less environmentally desirable containers, or too expensive to convert away from them in the short term.

International Issues

The GEE packaging component would be relatively simple to implement on a national basis. Once the taxes are in place, the market would take care of the rest. The question is, how do we ensure that local manufacturers and businesses are not put at an unfair advantage with the implementation of such a system?

Packaging does not represent a large percentage of the total value of the products we purchase. Furthermore, once markets are developed, used containers and those made from recycled materials may come out to be cheaper than even unlevied new ones. As such, there might not be a need to do anything.

However, imported packaging itself (empty boxes, bottles, etc.) would have an unfair competitive advantage. To ensure fairness, governments may choose to implement the levies that are applied to local packaging on all imports and optionally rebate them on exports. This could be done under either the double- or the individual-taxation scheme. Foreign manufacturers using recycled standard containers or countries having such programs could qualify for tax exemption through bilateral agreements.

A second option would be to apply a uniform but lower tax on all imports. This would offset some or most of the unfairness in competitiveness and keep things simpler.

A third option would be to charge levies at points of sales in stores. That would be much more bureaucratic, greatly multiplying the number of places from which governments would have to collect. However, it would have the benefit of treating both locally-made and imported products in the same way. The bureaucracy would shift from manufacturers to retailers. Such an approach could make it difficult to distinguish between new and reused containers.

In the long run, the solution may lie with international agreements since common standards would greatly simplify things.

Standardization

Taxes on packaging would shift markets to environmental alternatives. As already stated, glass is one of the best options as it is reusable. However, the fact that containers can be reused does not mean that they will be and not go straight to the landfill. Many fruit juice and drink bottles are not reusable. Others are never recycled despite the refundable-deposit system. Glass is a depletable resource.

The plethora of formats currently existing on the market makes it generally unprofitable to collect containers and process them for reuse. Unless special measures are taken to address the issue, we could still end up adding substantially to our garbage problem—not to speak of wasting a very useful resource—even if we have a sound recycling and reuse strategy. As such, countries serious about the environment should consider standardizing formats. Simple regulations could easily achieve that and result in broader reuse markets, higher rates of return, and savings.

The large variety of sizes and shapes currently in existence makes it difficult and often unprofitable to process containers for reuse. The market for each type is very small, and trying to collect, process, and warehouse them would be costly given the limited volumes. Even in the beer industry, there are dozens of bottle designs although it does not appear to be so. Most are very similar but specific to companies, fragmenting the market and making it more expensive for businesses to reuse them.

If containers were regulated into a minimum number of categories and designs, reuse could become more profitable and attractive to businesses of all sizes. Standardized sizes and formats would promote larger markets. For example, there would be a very few types of each of 5 ml, 10 ml, 50 ml, 125 ml, 250 ml, 350 ml, 500 ml, 1 liter, etc. jars and bottles. Design specifications would be available to both local and foreign container manufacturers so that overseas exporters who wish to qualify for a tax exemption could do so. A levy differential between non-standard and standard containers would be put into effect to promote the shift to the latter.

Standard containers would be slightly less expensive to produce than their non-standard equivalents as they would be manufactured and handled in larger runs. As most companies would purchase the same types of containers, their market size would increase and they could be more easily reused as the larger volumes would make it profitable to collect and resell them.

The processing, transporting, storing, and wholesaling of used standard bottles and jars would be much less costly as several companies within an area would share the same pool of containers, reducing inventory expenses and allowing their processing for reuse to be carried out locally. This would naturally promote a shift to them.

Furthermore, buying *standard* would mean buying green. As such, consumer preferences would likely shift towards this more environment-friendly alternative. Using standard containers would essentially be free publicity and a marketing advantage for most companies.

Under an individual-taxation scheme, governments could impose a higher levy on new standard containers—both local and foreign—to shift the market to used ones. Food processors and other companies would naturally gravitate towards the cheaper alternative, generating a demand for them, and essentially creating a reuse market.

Under a double-taxation scheme, the second levy would act to make new standard containers more expensive than used ones and promote a shift to the latter.

What would be the result of such an approach?   Store shelves would look different.   Groceries would be heavier to carry. Consumers would collect their standard used jars and bottles and sell them for cash to recycling firms that would process them for resale. Reuse would be achieved without the bureaucratic and inefficient need for deposits and refunds, or the involvement of governments.   An enormous amount of resources would be conserved. Landfill and intermediate chemical use would decrease drastically.

Used Container Processing

The beer industry often reuses its own bottles. It manages to do so because its product is pervasive in society and major players dominate the market. Large numbers of containers go back and forth between brewers and consumers. That allows for economies of scale to take place. Big companies have the capital to invest in bottle processing machinery (washers, sterilizers, etc.), but smaller ones often cannot afford it.

Many small towns would likely not be able to economically process used bottles and containers if packaging levies are implemented without standardization. In most cases, they would have to be shipped out of town to larger centers for processing. The fragmentation of the market would add to costs, and the diversity of formats would make it unprofitable to collect many types of containers.

Packaging levies would reduce the variety of designs as companies shift to containers in lower tax brackets. The market would become less fragmented. As a general rule, the more formats, the larger the markets need to be for profitability. With fewer ones, smaller centers could also have viable reuse processing operations.

Supply and Demand in the Reuse Market

Since the GEE strategy does not involve a deposit system, the price paid to consumers for their empties could vary depending on supply and demand and by localities. As such, the market could not always be relied upon to set a price which would ensure that recycling and reuse do occur.

Larger communities should reach high levels of efficiency and be the most beneficial to consumers. Smaller and less competitive ones may have to be supported by regulation. Governments may need to establish a minimum amount paid per empty container. It would have to be high enough to ensure that they are returned and that conservation strategies bear fruit.

A number of states in the U.S. have bottle return systems. The average deposit charged is about $0.05. Although some programs are relatively successful, the rate of return of others can be as low as 60%. Canada's deposit fees range from $0.05 to $0.40. To provide enough incentive for people to return most of their empty containers and achieve acceptable recycling standards in today's context, a minimum levy of $0.20 to $0.30 may have to be charged. In places where markets would not occur naturally and be profitable, municipalities would likely have to run recycling and reuse programs themselves. As in other GEE components, all of the above would be compensated for by a drop in other taxes.

The GEE would yield very tangible benefits in the packaging sector. A lot more would be reused and recycled than currently is. Waste would become valuable. Companies would actually compete over your garbage. Resource depletion and landfill expenses would be greatly reduced, intermediate chemical usage from the manufacturing of new containers would decrease, and the cost of recycling programs to municipalities, eliminated in many cases.

As packaging is massively used everyday, governments should not hesitate to use high standards of reusability, renewability, and biodegradability in its respect. They should do so urgently. The industry is in dire need of a comprehensive policy ensuring that reuse is maximized and that wastage is minimized. The only issue here is convenience, and our inability to get our act together.

GEE Management of Renewable Resources

Governments may also choose to tax some renewable resources to avoid over-exploitation. For example, new paper could be levied (at the producer level for collection efficiency) to reduce pressure on forests.

Recycled paper products would then become cheaper than new ones, promoting conservation and saving landfill space. A stronger demand for them would be created. Market forces would kick in, leading companies into the business of collecting paper and reselling it for profit to recyclers. Some of this is occurring at the moment but on a much smaller scale than it should.

The GEE could be pushed further into a full management strategy for renewable resources. To promote reforestation, governments could tax old-growth lumber in order to shift the demand to timber coming from land that has been replanted. It could assess higher levies on lumber produced through clear cutting or other poor management practices (assuming those are not regulated out).

The same strategy could be applied to fisheries, with poorer management approaches penalized by taxation, or species with dwindling stocks assessed levies to increase their price and decrease demand. Of course, all of this would also be done in a revenue-neutral way.

The GEE would be a powerful market-based mechanism to manage renewable sectors of the economy. It would provide for them the same benefits that it offers as a conservation strategy for non-renewable resources: simplicity, flexibility, market friendliness, minimal bureaucracy, etc.

The New Consumer Environment

Resource conservation would include a combination of taxes on non-renewable resources to decrease their use and create natural markets for recycling. A second set of levies targeted at non-renewable packaging would deter its use and reduce unnecessary wastage. Standardization and taxes on new containers would lay the foundation for a solid reuse industry without the need for the inefficient deposit-refund system.

Levies on specific renewable resources could also be used to promote conservation and prevent over-exploitation by giving a competitive advantage to recycled products. Most municipal recycling programs would be taken over by private entrepreneurs. The new market-based approach to resource conservation in the packaging sector would be much more efficient and effective than the current patchwork of taxpayer-funded programs. Again, remember that all these taxes would fill governments' coffers and that overall taxation would remain the same.

The GEE would slowly reshape the economy. Store shelves would take on an unusual appearance at first. Some types of foods may look funny in glass jars. Bulk sections in stores would likely expand. Most soft drinks would again be sold in glass bottles and be heavier to carry.

We would not have the diversity of containers we are used to, but that would only mean saving resources and the environment. We would still see the familiar logos of food companies on glass jars.

Flexibility and Scalability of the Market Approach

Choice would remain a component of a market-based resource conservation strategy. Non-standard, plastic, and most other types of containers would still be legal but more expensive. A soft drink company that does not care about the environment could continue to use plastic bottles, but its product would be pricier because of the tax on non-reusable containers. Likewise, consumers would be able to continue to buy unenvironmentally packaged goods but at a higher price.

Choice adds flexibility to the system. As already stated, it is an option that is often preferred by businesses as it gives them time to adapt. The flexibility of a market-based strategy would make it more acceptable to everybody. The approach is fully scalable, and the level of taxation as chosen by society would determine how much more one would have to pay for convenience.

Contaminants

The case for the reduction of pollutants has already been made many times in environmental literature. As such, the issue will not be rediscussed here in any detail. Contaminants would be a major target of a GEE strategy. Some issues—appropriate tax levels, international competitiveness, etc.—are very similar to those for non-renewable resources.

One of the differences with respect to contaminants is that they are more substitutable than metals. There are generally many alternatives available. Also, their contribution to the final cost of many products is often much smaller than that of metals. As such, their impact on the price of consumer goods would generally be less important. These are some of the considerations that would have to be examined closely in defining and determining policy.

Under the GEE, contaminants would be taxed to reduce their use and shift industries to cleaner substitutes. The determination of what should be levied and at what level is much beyond the scope of this book. Each chemical has its own properties and effects on the environment. Generally speaking, the guiding principle for those would be, the more toxic or harmful a compound, the higher the tax applied.

Industrial and Domestic Contaminants

In the industry, undesirable chemicals and other compounds would be taxed at the producer level. The byproducts of the manufacturing process are more problematic as they are often not visible to authorities. They are not bought like inputs but produced internally. They are not sold either, making them difficult to track. In some cases, levies might be appropriate, but in others regulations might have to be involved.

Many of the pollutants plaguing the planet today are not industrial. They are not intermediate chemicals or byproducts of the industry. They are the very goods produced for us consumers. They comprise the various household cleaners, laundry detergents, solvents, etc. we employ everyday. These and other harmful end products would also be targeted by the GEE (taxed at the manufacturing level) as they are used massively, in millions of households around the world. Relatively high levies should probably apply to them as many can be easily replaced by more environmental alternatives.

The Agricultural Sector

Ironically, the industry that puts food on our tables is also a very significant source of pollution. Modern intensive agriculture uses a variety of herbicides and pesticides that degrade our water systems and the environment. The GEE would and should target these contaminants to reduce their use and shift the industry and its R&D towards more organic alternatives and practices.

Close attention should also be paid to domestic herbicides and pesticides for the same reason. Some urban centers in North America have already made moves in that direction. These would fall under the contaminant strategy of the GEE and would be taxed at the manufacturing level.

A second source of pollution relating to modern agriculture is the widespread use of chemical fertilizers. They are responsible for the explosion in productivity that the industry saw in the last century but are also prime culprits in the degradation of our rivers and lakes as they promote the growth of algae that choke them and the fish that live in it. They would be targeted by the GEE to reduce their use and promote greener practices.

A GEE strategy in the agricultural sector—assuming a government opts for it—would raise the price of foods grown with chemicals and artificial fertilizers, shifting consumption to better and healthier organic alternatives. People would consume fewer contaminants, antibiotics, and carcinogens. Agricultural land and the environment would be protected. Again, everything would be revenue neutral.

12. The Fossil Fuel Sector

Energy will be explored in conjunction with the automobile industry. Both have a massive impact on the planet and are undergoing significant changes.

Oil and other fossil fuels are currently vital to just about all countries around the world. Some are dependent on them to fill their energy needs. Others—like many states in the Middle East—rely on them for income as net exporters.

Even within a country, world oil politics can have a tremendous impact. For example, most of the petroleum production in Canada currently comes from one part of the country, Alberta. Recent price hikes have meant billions of dollars in profits for that one province alone. A lot of that money came from the high prices charged for oil products to other Canadian provinces, some of which are significantly poorer than the already well-off Alberta.

The politics of fossil fuels are complex and far-reaching. What is for sure is that nobody is immune to them.

The Intermediate Phase

The oil industry is not about to disappear. Gasoline and diesel are still widely used around the world. As well, there is too much capital involved, and the lobbies are too big and too influential to be displaced. There are also many mega-projects currently under development around the world.

From an environmental perspective, it would be desirable to phase out fossil fuels as quickly as possible. However, because of the reasons mentioned above, most changes will occur gradually. An intermediate stage for the industry would probably lead to a stabilization in the use of fossil fuels in many countries as renewable energy alternatives are phased in. However, a global decrease in consumption might not occur given the continued growth of the world population and the emergence of massive markets like China and India.

What might change this is a stronger commitment to greenhouse gas reduction, but support for the idea continues to waver as the Copenhagen Summit in December 2009 showed. The only other hope at this point is the adoption of more powerful environmental strategies, especially those that would make it easier and more cost-effective for countries to reduce carbon emissions. The Green Economic Environment is one such approach.

The Market Approach to Energy

When the price of oil went up in the 1970s, many people and businesses bought into renewable energy technologies only to see their investment amount to nothing in the mid-1980s and 1990s when oil prices collapsed.

The GEE proposes to tax fossil fuels, such as petroleum and coal, for several purposes. Firstly, higher prices would reduce consumption, hence mitigate the production of greenhouse gases. Secondly, taxing fossil fuels at a level sufficiently high would make profitable the renewable energy industry and further the growth of a leading-edge sector of the economy.

The GEE would impose a variable levy on the price of a barrel of oil (either imported or locally produced) to raise it domestically to a set target, for example, $150.00. For a going market price of $120.00, the tax would add up to $30.00 and would be periodically readjusted to maintain the final price of oil at about $150.00 over time. This would create a predictable and stable environment that would not only foster the growth and development of alternative energy technologies but also prevent the crash of the new sector and the loss of precious green investment.

The GEE would be collected at the producer level. However, unlike other mineral levies, it would target domestic supplies and imports only. Exports would be exempt. This would raise the price of gasoline domestically and encourage conservation. But it would allow international market pricing mechanisms to continue to oper-ate for petroleum, a resource vital to all countries, many of which are struggling economically.

Remember that the GEE is revenue neutral. As the fuel levy is collected, income and sales taxes would be reduced in proportion.

Benefits of a GEE Fossil Fuel Strategy

The GEE fossil fuel strategy would yield many benefits. The simple tax on petroleum would in one fell swoop conserve resources, reduce pollution and greenhouse gases, promote the development of the renewable energy sector, and decrease depen-dency on the Middle East. The total benefits are so large and far-reaching that we should immediately move ahead with the strategy. The GEE fossil fuel approach is not the path to the future: it is the eight-lane highway to it. Alone, it would achieve the goals of the Kyoto Accord and yield multiple environmental benefits.

Currently, environmental strategies comprise most often a panoply of regulations and incentives such as grants and tax breaks, all of which require huge bureaucracies and a lot of monitoring. A tax on oil charged at the producer level where there are very few players would be much more efficient. With it, the law of supply and demand would on its own promote the development of the renewable energy sector without government intervention. This GEE strategy for fossil fuels would be efficient and produce a green energy growth sector that is stable and more competitive.

Fossil fuel taxes are never popular, but remember who pays for the grants and tax breaks used to fund environmental and renewable energy initiatives. We do. One way or the other, we pay. These expenses would disappear under a GEE system. In addition, with the GEE, a lot of money that would otherwise be spent on monitor-ing and new bureaucracies would be saved. The end result would be a more competitive and stronger renewable energy sector, and one that is fine-tuned to market laws.

Cap-and-trade is not bad in itself, but it is a flexibility compo-nent tagged on to regulatory limits. It would provide indirect support to the renewable energy sector in the form of a market for emission credits.   However, the trade aspect would not reduce carbon emissions and would be more complex and less cost-effective than the GEE, among other things.

Under the GEE, other fossil fuels would also be taxed to reduce their use and decrease greenhouse gas emissions and other pollutants.   The levies would prevent a shift from petroleum to other non-renewable types of energy which could be equally, if not more, damaging to the environment.   Lower taxation could be applied to cleaner or less carbon-intensive technologies if these ever come through.   For example, natural gas—a cleaner burning and over 50% less carbon-intensive fuel—could face lower levies.

The GEE for Oil Producers

Taxing petroleum is likely to be a very sensitive issue for producing countries.   They tend to be very protective of their massive oil wealth.   The GEE would not result in their giving away the revenue from these resources.

Internationally, the tax would not be imposed on exports, meaning that none of the profits from them would be lost.   For example, Alberta, Canada, would continue to sell its oil to the U.S. at international prices just as it does now.   Nationally, regular prices would also remain in effect for producers.   For example, Alberta would continue to sell its oil to Canadians as per normal supply and demand pricing although consumers themselves would pay a tax on top of it.

Local producers and importers of oil would remain on a level playing field as the levy would be collected from both.   Since the GEE is revenue neutral, consumers would also come out even, paying lower income and retail taxes in compensation.

Specific producing regions would not be cheated through the new system either.   For example, the levy that Albertans would pay to the federal government would not go to other provinces but would be received back as per the principle of revenue neutrality. The reverse would also be true; the levy that other provinces would pay the federal government would not go to Albertans but back to themselves.

What would change is that fossil fuel consumption would go down.   It would undeniably be good for the environment everywhere, including producer regions. Oil not sold now would only be preserved for the future. As such, the wealth from producer regions would not be lost or given away but a source of income for them over a longer period of time.

A Biofuel Strategy

Ethanol-blended fuels are gaining in popularity. Increasingly, big North-American car manufacturers are producing flexible-fuel vehicles (FFVs) capable of running on both gasoline and ethanol blends up to 85% in concentration.

The new fuels are cleaner burning and more carbon efficient than pure gasoline and diesel. Furthermore, they can use existing distribution infrastructure. They are also good for local economies, benefiting the agricultural industry.

A New Set of Problems

Over the years, the U.S. and Canada have subsidized their agricultural industries by the hundreds of millions of dollars because, among other things, of the low market price of some commodities. In Eastern Canada, potato crops were often destroyed to prevent an oversupply and a price collapse. Now, they could be turned into ethanol.

A biofuel strategy would be very positive in those sectors, and things have already begun to change. In the fall of 2005, France announced that it would turn part of its overly plentiful wine production into ethanol. Canadian and American farmers are already doing better from the rising demand for biofuels.

The shift to renewable energy would reduce waste and losses and generally make agriculture much healthier financially. The new economic landscape would eliminate the need to subsidize it with taxpayers' money. It would also result in cleaner burning fuels, lower greenhouse gas emissions, and a reduced dependency on oil.

Unfortunately, not everything is rosy about biofuels. Food prices are rising partly as a result of crops and land being diverted to the production of ethanol. Italians made headlines in 2008 for complaining about a sharp rise in the price of pasta. The United

Nations was talking of a potential food crisis in developing countries as people could no longer afford the skyrocketing price of rice following the high cost of oil in the summer of that year. World grain reserves were at historical lows. This was the result of not only the production of biofuels from edible crops but also rising fossil fuel prices and speculation.

The New Economic Reality

Biofuels do present a number of challenges. There is obviously the question of competition with food crops for arable land. But this does not have to occur. Alternative fuels can be generated from garbage, inedible plants grown in unused fields, and agricultural and industrial byproducts. Governments would therefore be able to intervene and exercise some control on the problem.

That would be important, especially in developing countries where many people can already barely afford food. The use of edible crops and good agricultural land for the production of biofuels can be regulated. It is a problem for which there is a solution.

A second issue is that of low net efficiency: it takes a lot of energy (often from gasoline and diesel) to produce biofuels. The issue should improve with research and technology. More importantly, the Green Economic Environment will make it possible for renewable energies to sort themselves out through markets. Because fossil fuels would be taxed, biofuel sources with low net efficiencies (i.e. requiring a lot of energy to be produced) would be less profitable and, as a result, attract less investment when not abandoned completely.

Note that electricity generation from hydro, wind, tidal, geothermal, and solar sources is not expected to affect the price of food. It does not compete with it for land. As such, not all new alternatives would cause problems. Electric cars have begun to hit the roads and are increasingly becoming a viable option for the future. That will decrease the demand for ethanol.

Biofuels and other sources of renewable energy will be an evolving landscape for some time. We should not panic as new problems and challenges emerge. Most will be solved.

New Land for Agriculture

Can new land be developed for the production of energy? Areas that are valuable to us for one reason or another (old-growth forests, wildlife habitats, etc.) should be protected, but could other types of land be converted and have a positive effect in terms of carbon reduction?

Large tracks of tropical forests are currently being cleared and planted with palm trees for the production of vegetal oil for biodiesel. Wetlands and grasslands are also being targeted for conversion to biofuel production. Unfortunately, these are already carbon sinks (i.e. areas that store significant amounts of carbon). As such, many argue that their clearing adds to global warming problems (Common, 2008, January 25).

The real question is, would we be making the same mistake as we did before by trying to create new land for agriculture: treating the symptoms rather than the disease? Clearing strategies would lead to the progressive *denaturalization* of the planet by the replacement of forests, wetlands, etc. with crop fields. Is this really the answer, or is it the problem?

What put us into this mess in the first place is unbridled consumption and population growth. A permanent solution to environmental problems will need to address these issues.

The new economics of the environment will have to be closely monitored and regulated. The science will evolve as we go along, and governments will find out what works and what does not. Efficiencies will likely improve over time and with economies of scale. The new green environment would certainly create new challenges and call for changes in laws and regulations.

In the long run, food prices will probably trend higher not because of any biofuel strategy but because the cost of oil will keep increasing. This is just the new reality of resource depletion, a reality that we have only begun to get a taste of. The GEE would slow down that process by promoting alternative sources of energy, hence preserve oil reserves and slow down the increase in price of the commodity. In addition, it would favor green goods and make them cheaper. And that includes food.

A Market-Based Biofuel Strategy

Could governments use regulations to force the addition of 10% ethanol or more to gasoline and promote a transition to renewable energy that way? Technically, yes. However, the move would cause problems in that the oil industry would have to purchase quite suddenly a large amount of ethanol for which there is little supply at the moment. This could result in high costs for the fuel and chaotic gasoline prices.

The best approach could include regulations, especially initially, but should be based primarily on market forces. A high and stable oil price would make ethanol blends and biodiesel more competitive, which would promote a natural switch to these cleaner and more environmental options without creating a supply crisis. The higher the target price, the more competitive biofuels would become. As such, governments would be able to implement a strategy progressively and control the speed of the process.

A market approach to bioenergy would ensure a gradual and smoother transition by allowing production to grow and respond to demand as opposed to using regulations and creating a supply crisis.

A biofuel strategy would create markets, sustain crop prices, and prevent the loss of investment and jobs. The growth of the industry would ensure a thriving agricultural sector for the future and help redistribute wealth regionally.

Hydrogen: Panacea or Illusion?

Hydrogen was the focus of much media attention a couple of years ago. This potential energy of the future captured the imagination of the public because its combustion or use in fuel cells to create electricity is essentially pollution free, water vapor being the only exhaust emission.

The main reason why hydrogen seems to have overtaken electricity in the race for clean power in transportation is that it provides a longer driving range. The best electrical storage technology (batteries) to date could satisfy urban commuting needs but remains impractical for longer range or heavy transportation.

There are a number of problems with hydrogen. Firstly, the gas is highly explosive. Safe storage technology is under development, but costs are high and may remain an issue. A second concern is that it is only an intermediate fuel. There is no source of hydrogen per se. There are no vast pools of it underground as there are for oil. It needs to be produced from or with other sources of energy, for example, coal, natural gas, etc. An obvious issue is that producing hydrogen from fossil fuels would release in the atmosphere large quantities of carbon dioxide, a greenhouse gas. As such, it would not be an alternative or a renewable energy.

Hydrogen can be produced with electricity through a process called electrolysis. However, each time you transform one energy into another, there is not only the cost of doing so but also a conversion loss. 100 units of electrical power could yield only 70 of another energy. As such, it is better to use it directly rather than convert it into something and then back into electricity in the fuel cell of a car. Different avenues are being explored for hydrogen production, but it is unclear at this point in time whether or not the gas will become the energy of the future.

A third concern is the massive amount of capital investment a conversion to hydrogen fuel would require. Simple, well understood, and cheap infrastructure already exists for electricity: knowledge, distribution lines, motors, etc. Such is not the case for hydrogen. Capital expenditure for a hydrogen society would be high and probably fairly risky as a lot of government planning would be involved. It would require a huge commitment and a leap of faith into a future that may never happen.

A fourth concern is the old chicken or egg problem. Which came first? Do you build a huge and expensive distribution system first, hoping that people will switch to hydrogen transportation? Or do you build cars for which there is yet no fuel distribution system?

Breakthroughs may change the odds for hydrogen, but currently there are still too many questions left unanswered. An educated guess at this time points to limited applications and niche markets.

A large-scale future for this fuel may come through if research leads to ways to convert the world's massive coal resources to the gas without carbon emissions or pollution. But again, even the large reserves of the fossil fuel will also eventually run out and its mining is a significant source of pollution. These are aspects that will also need to be considered for hydrogen.

Compared to the renewable energy sector, the fossil fuel industry is highly concentrated. Its immediate and long-term interests are not necessarily the same as those of society as a whole. Global warming problems preclude our continuing down the fossil fuel road. Diversifying into decentralized renewable energies would spread wealth around and benefit future generations.

Possible Breakthroughs for Electricity

For a few years now, EEStor (a U.S. company) has been working on a super battery, the EESU (Electrical Energy Storage Unit). The ultracapacitor would triple the current driving range of electric cars to about 300 miles (500 kilometers) and would recharge in minutes. This would surpass by a long shot lithium-ion batteries and the best technology available to date.

The battery would be nontoxic and apparently would not degrade over time from recharging like others on the market today. This would make it more conservational and less harmful to the environment. EEStor also reports that its self-discharge rate is very low and that it would perform well at low temperatures—unlike other batteries—making electric vehicle transportation much more viable in colder climates.

Zenn Motor Company, a Canadian manufacturer of electric vehicles, has heavily invested in the technology and the EEStor company. It is the only automobile manufacturer with a license for the use of the ultracapacitor in cars. It had originally planned to put out vehicles with EESU batteries at the end of 2009, but that schedule has been postponed. 2010 should be the do or die year for the EESU.

Skeptics believe that such a technological leap is impossible and warn about the dangers of meltdowns for the high-energy battery in car accidents. The technology has had successful initial testing in independent labs, but the final word is not out yet.

What is certain is that if the EESU does come through, it would change the whole picture for the transportation industry. Electric cars would then be able to replace gasoline vehicles without the usual inconveniences. This would likely kill the market for both the hydrogen and hybrid technology and would revolutionize transportation and our urban environments.

Several companies are looking into improving lithium batteries. One of a number of interesting developments is the use of lithium iron phosphate to speed up recharge time by up to 100 folds (MIT).

Others focus on increasing energy storage capacity, among them single crystals of lithium cobalt oxide (Toyota), silicon nanotube anodes (Stanford University, Hanyang University in Korea, LG Chem), lithium-air technology (IBM), and ionic (U.S. Department of Energy, Scottsdale) and lithium-sulphur batteries (NSERC-funded lab, University Of Waterloo, Canada). These new technologies claim to be able to increase energy storage capacity by three to ten times and would also change transportation as we know it.

We will see many other advances in the near future, some of which should turn out to be very significant. The automobile industry's recent surge in interest in alternative cars will result in a lot more money going towards battery research. However, the technology will also be crucial for the production of electricity at home, enabling the storage of intermittent energies (wind, solar, etc.) while increasing the supply for a rapidly growing electric transportation sector.

As some have already pointed out, millions of batteries in cars would provide a way to store large amounts of excess grid electricity. While it is possible for governments or power companies to orchestrate a system for this, the GEE would make it happen on its own through market forces.

Companies would simply have to offer cheaper rates at peak-production or low-consumption times—which they already do, excess energy currently being sold at lower rates to companies that require a lot of cheap electricity, for example, aluminum producers.

At times of excess production (e.g., on windy and sunny days) or low demand (for example, at night), special meters would simply turn on certain outlets and let home and car batteries be charged with low-cost energy. The cheaper electricity would simply be used at peak consumption times at home or for transportation. The rest could be fed back into the grid.

13. The Automobile Industry

A Blueprint for a Renewable Energy Future

Doing a detailed study of renewable energies is beyond the scope of this book. Rather, this section will try to identify major trends and highlight the issues and factors that may enable us to select policies for the short term and develop a blueprint for the energy and transportation sectors of the future.

Special attention will be paid to strategies that may benefit us on several counts as opposed to those furthering a single goal. Policies that may be able to bridge us to the medium-term will also be given preference as it makes no sense to develop plans for an infrastructure that will become obsolete in a decade. Decisions regarding a future for energy and transportation may ultimately be a matter of social choices as the various options open to us have different implications.

The Fuel-Cell Future

At the moment, there are two major trends in the future of transportation and energy. The first one, fuel cells, is still at an early stage of development and depends on new technologies. Hydrogen, which sparked the idea of clean transportation, would necessitate a huge investment in infrastructure and has fallen out of favor for this and other reasons. While the wealthier parts of the world may be able to afford such a strategy, most countries may find it simply too expensive or not the most cost-efficient option for them.

However, new ceramic fuel cells have significantly increased the efficiency of making electricity from natural gas (NG). The process would produce carbon dioxide but only about half as much as gasoline. NG is also much cleaner than other fossil fuels in terms of toxic contaminants such as sulfur dioxide and nitrous oxide. It is abundant, widely distributed around the world, and currently significantly cheaper than petroleum. As well, the economics of its efficiency in transportation are very promising (Blakeslee, T., 2009, September 23).

Natural gas is mostly methane, which can be produced or captured from biomass such as garbage, sewage, manure, crop residues, etc. Biogas, as it is called, is renewable and can use the existing infrastructure for NG distribution.

Biomass only degrades into methane under anaerobic fermentation, that is, without exposure to oxygen. A few apples under a tree will not produce biogas, but fermenting them in a sealed barrel would generate methane and be carbon neutral.

Landfill sites, sewage lagoons, manure tanks, and large piles of organic matter with limited oxygen penetration will naturally emit methane. As the latter is a greenhouse gas 20 times more potent than carbon dioxide, capturing it from biomass (i.e. not letting it be released into the atmosphere)—as opposed to producing it—would actually be carbon negative and have a reduction effect on global warming several times that of CO_2 for an equal weight.

The combination of the wide availability, low costs, and shared infrastructure of NG and biogas as well as the renewability and potential for decreasing global warming pressures of the latter makes methane fuel cells a strong contender for the future of transportation. The technology already exists as small-scale 2 kW stand-alone power generation units but has yet to be adapted for use in cars. Methane fuel cells could be a very significant avenue for the future of transportation.

The Biofuel-Electricity Future

The biofuel-electricity future is a mix of both old and new technologies. Research will lead to exciting new developments and improve production techniques and hardware, but some of the science behind it is quite old. Vegetable oils—from which biodiesel can be derived—have been produced for a long time. Ethanol—drinking alcohol—is actually something that has been around for millenniums. Regular gasoline engines can burn 10% blends without modifications. Ethanol production is relatively simple and well understood and would be of prime benefit to the agricultural industry, at least in the U.S. and Canada.

The future of biofuels would entail low infrastructural spending as a lot of the technology is already here and can use existing distribution systems. Flexible-fuel vehicles increasingly represent a growing market in North America.

Brazil has already tested and proven the feasibility of a large-scale implementation of a biofuel strategy. The South American country already has millions of cars running on ethanol produced from local sugar cane crops—with very positive results for its economy.

A biofuel-electricity strategy would rely on ethanol and biodiesel for long-range transportation. Urban commuting would be based on electric vehicles. At the current state of technology, their shorter range (about 150 miles or 240 km in better models) does not make them a realistic option for long-distance traveling, but their lack of emissions makes them perfect and as clean as hydrogen for commuting to work.

An electricity-based urban transportation strategy would greatly decrease pollution and smog problems. It would also reduce the need for biofuels, whose production competes for agricultural land and causes increases in the price of food.

Looking for Common Ground

Future technological directions are paramount in considering options for long-term strategies. Optimally, current policy choices should support future options whatever they may be. An interim strategy should yield an infrastructure that would be able to support both a fuel-cell and a biofuel-electricity future.

Currently, world energy is derived from a variety of sources: oil, natural gas, coal, hydro, etc. This is for good reasons: needs are so massive that concentrating on only one kind would result in a supply shortage and skyrocketing prices. In addition, some types of energy are better suited for certain applications. As such, we will have to continue to get energy from several different sources in the future, and the more options are open to us, the better. To do this, we have to look for common ground between the fuel-cell and biofuel-electricity scenarios.

The Convertible Electric Vehicle

A fuel-cell car is actually an electric vehicle powered by a cell as opposed to a set of batteries. As such, both could share the same automobile architecture with the exception of the energy module. Their motors and other components could be exactly the same.

Designing and developing convertible electric vehicles may be the solution to the *chicken or egg* problem posed by a monolithic hydrogen strategy or to make other new technologies adopted earlier. Cars using either batteries or fuel cells as a source of energy (and convertible from one to the other by the replacement of the power module) could significantly speed up changes in the automobile industry and prevent investments from becoming obsolete in case one technology does not come through or falls out of favor. Convertibility would allow both manufacturers and consumers to switch from one to the other if the price of one type of energy rises significantly.

The Electrical Grid

The power grid would essentially be the only infrastructure needed to support electric vehicle transportation in urban settings. Batteries could be recharged at off-peak times (either at home or work). The grid is already a common infrastructure for electricity produced from a number of sources: hydro, coal, nuclear, etc.

The same network could support many of the future's renewable energies. In fact, it has already begun to happen. Governments are increasingly talking about new policies allowing the buy back of surplus electricity from households. If there is a significant shift in transportation from diesel and gasoline to electricity, the consumption of the latter will increase and drive prices up. New sources will be needed to meet the rising demand and keep costs down.

As such, electrical networks would become central in increasing supply. The grid is already providing support for many wind farms. With little additional investment, it could also support millions of micro-producers—for example, anyone purchasing a backyard wind turbine or solar panels with the intent of selling surplus energy into the grid. This has already started to happen in many countries. With increased demand, micro-producers would become an important source of renewable energy supply. Wind turbines and solar panels may just become a ubiquitous part of the landscape in the near future.

The grid would become central to not only increasing the supply and delivery of renewable energies but also supporting hydrogen without the need for massive infrastructural investment. The clean fuel could be produced at home from grid electricity in small electrolysis machines. At the moment, efficiencies are not great, but that may change in the future.

Direct home production would remove brokers and retailers from the equation, meaning that it could favorably compete with commercial ventures. It could also take advantage of off-peak rates, which would also serve to lower costs. Although hydrogen has fallen out of favor, the future might still hold some promises in this respect.

The electrical grid is a common infrastructure that is already in place and offers the possibility to significantly increase energy supplies and do so from renewable sources.

The Automobile Industry Under the GEE

The Green Economic Environment strategy proposed in this book would make most of the above happen on its own. It would promote renewable energies and provide stability for the sector, preventing the loss of investment—as has occurred when the price of oil dropped—and ensuring growth for the future.

Implementing a biofuel strategy would be relatively simple. All that is needed is for oil prices to be high enough to make renewable energies competitive (as discussed in the chapter on fossil fuels). The market would take care of much of the rest. It would make renewable energy and ethanol blends cheaper than gasoline and diesel, and ensure that fuels that are carbon intensive (e.g. ethanol produced from food crops) are more expensive and fall out of favor.

Carbon-negative technologies such as *methane capture* (but not production) could be the object of further promotion under the GEE because of its multi-fold impact on global warming from the removal of a potent greenhouse gas from the atmosphere. The strategy could be as gradual or as fast as we would want and essentially without quotas or regulations.

There are two main sectors in transportation. Each is qualitatively different in terms of needs, challenges, and markets. Long-range transportation, such as the trucking industry and holiday traveling, requires vehicles capable of going over long distances without refuelling. Electric vehicles are not currently appropriate for long-range and heavy freight and would not be suitable alternatives for these purposes at this point in time. As such, long-distance transportation would continue to be based on gasoline, diesel, and blends that include renewable fuels.

Urban commuting, the second main sector, is characterized by a multitude of smaller vehicles operating within densely populated areas. These play an important role in urban air pollution and smog problems. As such, everything should be done to make their emissions as clean and pollution free as possible. The sector would be an ideal candidate for electric or fuel-cell vehicles.

Transition in the Automobile Industry

Both the fuel-cell and the biofuel-electricity future would find a common ground in a convertible electric vehicle. Governments and industry could work together in order to design a basic frame for modular electric vehicles that could be powered by either batteries or fuel cells. These would initially be operated with the former, making them highly suitable for urban commuting. They would immediately provide cleaner urban environments and decrease fossil fuel consumption and greenhouse gas emissions.

When the fuel-cell technology comes through, the same modular vehicles could be used. These commuter cars would provide continued work in and a transition for the industry. This would mean a cleaner environment now, create work for the automobile industry, and provide a transition to both battery and fuel-cell technology.

Convertibility between electricity and fuel cells would allow people to switch easily between different types of energy according to supply and cost. This would support a better and more stable price environment, provide a flexible and diversified energy strategy for the future, and prevent our being held hostage to fossil fuels ever again.

The strategy would conserve an enormous amount of non-renewable resources as an entire generation of vehicles could be upgraded to better technology—as is likely to happen in a leading-edge sector of the economy—as opposed to being scrapped and added to landfills.

From a business point of view, convertibility would provide the egg to the *chicken or egg* problem of hydrogen. Ten years from now, battery-operated fuel-cell-compatible vehicles would already be in wide use. Infrastructural investment would be less risky and could be provided by the private sector as opposed to taxpayers and governments. Of course, that assumes that hydrogen is not dead already.

The immediate implementation of a convertible electric vehicle strategy for commuting would revolutionize the urban environment. Cities would quickly become much cleaner and quieter. Smog would be significantly reduced or may disappear altogether.

During the transition period, we would live in cities where single people would use mass transit or electric cars to commute to work and rent a hybrid vehicle for holidays. Couples with two gasoline automobiles might keep one to commute to work and use for family holidays. The second one would be replaced by an electric car.

Appropriate electrical grid policies would have to be implemented alongside an electric vehicle strategy in order to increase the supply of electricity and promote renewable sources of energy.

14. Transportation

This chapter will explore in more details future options relating to the automobile industry, a very important sector in the economy of many countries and one that will see plenty of changes.

In the short and the medium term, the GEE would increase the demand for more environmental and conservational generations of vehicles. In the long term, two trends would develop as a result of taxation on non-renewable resources.

Firstly, manufacturers would begin downsizing cars. Secondly, we would see part of the production shift to remanufacturing. Automobiles would be kept for longer periods of time, repaired, and upgraded as opposed to being bought new. Greater standardization and increased modularity of architectures would make it easier for parts to be replaced and reconditioned.

These two avenues of development would spell a significant decrease in resource depletion and major changes in our cities and living environments.

The Hybrid Question

What about hybrid vehicles, those having both a fuel engine and an electric motor? They are often viewed as being the solution to all problems in the automobile industry of the future. Undeniably, they are a step forward. However, they raise several issues.

Early models really disappointed in terms of improved fuel consumption. There has been much improvement in the technology since, but they may fall short of how far the transportation of the future needs to go. As well, hybrids use up more resources as they call for both a fuel engine and an electric motor and are heavier and more metal intensive as a result. This will probably limit their future.

Hybrids might retain a place as family or holiday car but do not go far enough in terms of fuel efficiency and resource conservation as far as urban commuting is concerned. We can do a lot better. In the long term, they might be able to carve themselves a share of the market if fuel efficiency increases significantly and competing technologies like the electric car do not see significant advances. Because there is so much in the pipeline in that respect at the moment, the odds are against the hybrid.

For the last 20 years, there have been many calls for better public transportation in order to mitigate environmental problems. While the approach can be effective in large cities, it is not the answer everywhere or to everything. The reality is that personal vehicles are here to stay and the automobile will remain pervasive in society. Public transit has to be improved, but more effective and conservational vehicles have to be designed for individual transportation.

Public Transportation

Mass transit is an important alternative that offers multiple benefits to society. It reduces traffic and its related problems, among them, noise and air pollution. Most larger cities today would grind to a halt without it. Public transportation reduces the demand for fossil fuels and promotes the conservation of non-renewable resources.

It prevents the use and purchase of millions of vehicles. Moreover, buses, trains, and subway cars are built to last much longer than regular automobiles. Because of their initial price tags, public transit vehicles also tend to be repaired and refurbished more extensively. That makes them highly conservational, several times more than current automobiles.

There are different mass transit models currently in effect in cities around the world. Some involve standard fares regardless of the distance traveled. Others based on a concentric zoning system expanding away from city centers. Each has a certain amount of built-in inefficiency. For instance, the standard-fare system charges the same price to people traveling short distances as it does to those transiting much longer ones. The zone model addresses this by setting fares generally based on the distance traveled from city centers but does not reflect the intensity of travel routes.

In most cities, some transportation lines are heavily used while others are not. The buses, trains, or subway cars servicing the latter often run half empty or worse. That is not good. Despite the longer distances involved, major routes can be far more efficient than shorter ones because vehicles are in full use. They generate more revenue, and the actual cost per person—and to the environment— can be much lower than what passengers are actually charged.

A third model better reflects actual costs and is also more environmental. The zones of the second option are modified into a more organic artery system. The efficiency pattern of public transportation systems is brachial just like a tree, with trunks and major branches being highly efficient and smaller ones being less so. The artery model would make many long-distance routes cheaper and encourage people living in suburbs to leave their cars at home and reduce pollution.

Instead of being concentric, transit zones would extend like fingers from the central business district along main arteries. Fares would be cheaper on primary routes as these would be more extensively used. They would increase on secondary and tertiary lines.

Public transportation is a clear avenue for the future and deserves government support. It has many limitations, such as availability in suburbs and smaller towns, frequency of service, and practicality in certain situations. Government support to increase its use and benefits would mitigate many of its limitations; more people riding would mean more frequent service and route expansions.

As public transit will not satisfy all the transportation needs of the future, other ways of improving traffic, reducing pollution, and conserving resources have to be explored.

Cycling and Car Pooling

One of the most efficient means of transportation is the bicycle. It is heavily used in a number of Asian countries. Unfortunately, it is increasingly being replaced by scooters, motorcycles, and automobiles—with disastrous results for both the urban and natural environments.

Bicycles are not as popular in advanced countries although they provide a cheap, environment-friendly, and healthy alternative in terms of exercise. Many cities lack appropriate paths for them, making their use dangerous. The bicycle would do better under the GEE and would gain from being promoted by governments, be it in the form of paths, safety measures, or financial incentives.

Car pooling is also a very good and growing environmental option considering that most daily commutes are done by single persons in four-passenger cars.

Individual Transportation

Undeniably, the conservation of non-renewable resources would mean that automobiles would be kept longer and repaired more extensively. Many jobs would eventually shift from the new car industry to the parts and repair sector as well as pre-owned vehicle retailing. Refurbishing used automobiles and upgrading power modules would become a growth industry in the longer term.

Increasingly, cars would switch from being a disposable good with a seven or eight-year lifespan to something that is fixed, upgraded, and refurbished for two to three decades. New cars would cost more, but their resale value would also be higher.

Obviously, less metal would be used in designs. Vehicles would be smaller, lighter, and R&D would shift towards substitutes for metals: plastics, fiberglass, carbon fiber, composites, and biomaterials. The industry would focus on longer lasting and higher quality vehicles.

It would also move towards more easily reusable uniform chassis (where most of the metal in a car is located) and modular designs. Cars would look different on the outside but would be built on more similar basic architectures and with larger numbers of standard components to extend their lifespan and reusability.

Conservational and Environmental Cars

Although electric vehicles are not yet suitable for long-range transportation, the technology is essentially ready for the urban environment. Prototypes were built over 50 years ago. At this point in time, the best battery technology still leaves us wanting in terms of long-range transportation, but it is sufficient for urban and work commuting—which account for most of the driving that people do in a year. This does not represent a niche but the largest part of the market.

Since the electricity distribution infrastructure is already in place, cars would simply be recharged at home from a regular power outlet, often taking advantage of cheaper off-peak energy rates at night. That would change the economics of electricity.

Batteries are currently an issue in northern climates. A combination of better insulation and additional infrastructure could help mitigate the problem. In the Canadian Prairies where winters can be bitter, parking lots are supplied with electrical outlets so that cars can be plugged in even while people are at work. That could also be part of the solution. The additional infrastructure would represent some expense, but all components are mass-produced, can be quickly installed, and require little maintenance.

With the exception of batteries, electrical technology for cars is low cost because it is already well established and mass-produced. Maintenance for motors is also much less expensive than for combustion engines as there are no carburetors, radiators, oil changes, etc. That would also mean a lot less metal, hence a much lower initial price under the GEE system. As electric cars are not currently mass-produced, they are bound to be more expensive than they will be in the future. Their simpler and lighter technology should eventually make them significantly cheaper than their gasoline alternative.

Conservational and environmental cars could be designed and mass-produced relatively quickly if governments and industry cooperated to speed up the process. Cross-industry regulations could bring in more uniform chassis for longer lifespans, maximize the use of standard parts, and establish modularity to enhance repairability and allow for easy future fuel-cell conversion.

Government involvement could bring about a fair amount of synergy, leading to cooperation within industries, reducing financial risks, and decreasing the potential for the loss of investments.

Both modularity and convertibility would also prevent a large amount of resources from being invested in obsolescence.

At the moment, the incentive to move to clean-powered cars is growing. The GEE would increase the demand for greener automobiles and usher in new generations of conservational (smaller, less metal intensive, built to last) and environmental (powered by clean and renewable energy) vehicles.

SPV Transportation: Size Does Matter

We have now progressed from gas guzzlers to convertible electric cars. The GEE would take us one step further to the *single-passenger electric vehicle* (SPV).

A four-passenger automobile is not needed to transport only one person. Significant amounts of resources and energy would be saved in designing one-passenger cars for work commutes. The benefits of one-seater vehicles are many: smaller automobiles are more maneuverable in congested traffic, easier to park, and generally less costly to drive. Single-passenger vehicles would cost less, use up less non-renewable resources in their manufacturing, and probably be more than twice as energy efficient. Lower weight would also extend their driving ranges.

There is a market for SPVs as well as price, conservational, environmental, and traffic incentives to minimize the size of cars. The new single-passenger vehicles would be not only shorter, allowing twice as many cars in a traffic lane, but also thinner, allowing two vehicles to drive side-by-side within it.

At moderate and high commuting speeds, SPVs would run staggered on different sides of four-passenger car lanes, allowing twice the number of vehicles within a given space as currently while respecting safe driving distances and easing pressures on circulation. As they slow down to approach intersections, stop signs, and red lights, or are caught in traffic jams—where their congestion-reduction ability would matter most—two SPVs should be able to run side-by-side within current lanes, a given space packing in as much as three or four times the number of vehicles it now does.

Single-passenger conservational and environmental cars would nearly quadruple the current traffic capacity of roads. This could eliminate most of the circulation problems that plague commuters on a daily basis in most large cities around the world. Of course, it assumes that we do not increase the total number of cars on the road. The GEE would not lead to that if appropriate tax rates are set. It would make cars generally more expensive, preventing their proliferation.

In fact, increasing the number of vehicles on the road would defeat the purpose of designing more conservational cars and result in even more non-renewable resources being consumed. It would also undermine environmental alternatives such as public transit, car pooling, and cycling.

On their own, SPVs would yield significant conservational and environmental benefits. Reducing the total number of vehicles on the planet could further increase environmental gains but is likely to be politically unpopular. In the short term, maintaining the status quo in terms of number of cars on the road is probably the best policy as it would allow governments to bring in the GEE with much less resistance on the part of the general public.

Fourth Wave urban environments will be drastically different. SPVs will mean clean (emission free), quiet (electric motors can hardly be heard), and, in many instances, traffic-jam free transportation in urban environments.

Technical Issues Relating to SPVs

There are a number of technical issues relating to SPV transportation. Car width would have to be restricted if two of them are to fit within a single lane. Vehicles would have to be properly designed to prevent rollovers upon turning. Architectures would have to include such things as swivel technology and low centers of gravity. Car speed may have to be limited. Alternatively, cities could choose to redraw some traffic lanes.

Trends for the Future

GEE-based transportation would call for much higher quality vehicles, ones that would be built to last for a long time and be more repairable and upgradable. Some parts are already fairly standard in today's cars—wheels, tires, batteries, mufflers, etc.—and components like brakes can be refurbished. The used car industry already exists. As such, the GEE would not call for extreme changes, but it would refocus the sector.

Although advances will continue to be made, we already have the technology necessary for electric vehicles. What is currently missing is a market. The green economic environment proposed here would create one.

How farfetched is all of this? Since the first edition of this book was published, many of the things discussed in it have started to happen. There are not many single-passenger vehicles on the market at this point in time. However, the electric automobile is in and sizes are decreasing. The industry is increasingly talking about *biocars*, vehicles having parts made with bioplastics and composites produced from soy, wheat, canola, or sugar cane.

Some companies have already developed plant-based polyurethane foam for seat cushions. Volvo claims to use renewable materials in dozens of its car parts. Mercedes S-Class vehicles are also going green, their bio-components tipping the scale at 43 kg per unit (Stauffer, 2008, February 15). SPVs are not here yet, but the automobile industry is definitely moving into greener fields.

In the second half of 2008, the auto sector saw tremendous changes. SUV sales dropped sharply. Major manufacturing companies shocked analysts by suddenly deciding to close several SUV manufacturing plants in the U.S. and Canada. Most automakers started talking about either developing or producing electric cars and more energy-efficient vehicles. Some have models already on the market.

The Ready Market for SPVs

There is an existing market for single-passenger electric cars: families that already have two vehicles. Two multi-passenger long-distance fossil fuel cars or Sport Utility Vehicles (SUVs) are not needed for simple urban commuting. The second one—if it is really needed—could easily be an SPV.

Currently, most cars making the daily work commute are four-passenger vehicles which convey only one person. This way of getting around is highly inefficient not only from a fuel perspective but also from a non-renewable resource point of view. Two-car families are a ready market for SPVs.

Other possible buyers for the one-seater electric vehicle are single people. Instead of purchasing a gasoline automobile, many of them would choose to buy a lower cost electric car for commuting to work or school, especially once the GEE makes vehicles pricier. They would rent a gasoline automobile or a hybrid once in a while as necessary.

Remember that the GEE would be revenue neutral, and that while cars would be more expensive, people would have more money to spend.

Safety for All

The minivan and SUV markets really took off on the safety issue: the bigger, the safer. However, the real question is, safer for whom? Although they offer more protection to their own drivers, they may not be better for the rest of us. Much bigger vehicles are more dangerous to both other drivers and pedestrians, at least in theory. Greater safety for all lies in decreasing the average size of vehicles, not the opposite.

A lack of safety does not stop pedestrians from crossing the streets or many people from riding bicycles and motorbikes—which do not provide much protection in collisions with cars. As such, a market for SPVs will develop regardless of the safety issue.

Smaller cars are often thought to be less safe for their own drivers. This is likely true to some extent. However, design is a large component of safety. Race cars are smaller, yet they provide higher protection than your average automobile. Properly engineered SPVs could offer a reasonable amount of protection. They would also provide much safer city streets for everybody.

Fast-Tracking SPV Transportation

SPV transportation would bring in vehicles about 65% smaller than current four-passenger cars. That would mean, in theory at least, a 65% conservation of non-renewable resources—metals—a 65% reduction in intermediate chemical use, and at least a doubling of fuel efficiency.

Much smaller, resource-efficient, and more repairable one-passenger vehicles would offer a number of cost savings. This should initially translate directly into greater affordability compared to regular cars, especially under the Green Economic Environment. Mass production would give SPV transportation an even stronger competitive edge and, in doing so, achieve massive conservational, environmental, and traffic benefits.

To speed up the development of the industry, governments and automobile manufacturers could get together to design a uniform chassis for a one-passenger car that would be mass-produced and used as common architecture for all manufacturers in their first models. Cooperation would lower development costs, enhance modularity and reusability of chassis, and enable mass production even for the first models.

Pooling R&D resources would force on the industry efficiencies that would be beneficial to both consumers and manufacturers. Standardization could also occur internationally. Modular designs would create a platform that is both conservational and environmental and do so on a massive scale, worldwide.

GEE taxation on non-renewable resources and stable high oil prices would be the basis that would provide for steady and predictable growth of not only the renewable energy sector but also SPV transportation. Appropriate government support and industry cooperation would eliminate the high development risks and insecure markets for car manufacturers.

The consumer would be happy, the industry would be happy, jobs in the new automobile sector would be preserved, resources would be saved from building much smaller vehicles, and energy efficiency would improve.

The demand for electricity—which has the potential for being produced cleanly—would increase. The markets for wind turbines, solar panels, and other renewable energy technologies would take off. Micro-producers would join in the frenzy, selling excess electricity back to the grid and improving their own country's balance sheet. The shift to electric vehicles in urban transportation would decrease the demand for biofuels and lower pressure on food prices, perhaps averting widespread instability and a world hunger crisis.

SPV transportation would mean massive gains in resource conservation and energy efficiency. Those would occur to a large extent in a market-friendly manner. As necessary, regulations and incentives—including tax breaks and rebates on insurance—could be used initially to enable a quick takeoff for SPV transportation.

International Markets

Developed Countries

A huge market for SPVs would be in developed countries where they could provide an alternative to many of the cars on the road today. Together with double-passenger vehicles based on the same conservational and environmental technology (DPVs), they could take over the bulk of the market for cars.

Developing Countries

Most countries around the world cannot afford a North American style of transportation. Neither can the planet. The question of transportation in the developing world is highly complex. On the one hand, a proliferation of cars around the globe would be disastrous. On the other, the automobile is of great appeal to the growing middle class of many countries.

A significant part of transportation in the developing world is based on smaller vehicles: gasoline rickshaws, motorcycles, scooters, etc. A switch from fossil-fuel to electricity-based transportation in their case would be a major improvement. Smog and greenhouse gas emissions would be significantly reduced. The demand for cleaner domestically-produced renewable energy would increase, creating jobs in the local economy.

A properly implemented GEE scheme would lead to gasoline rickshaws, motorcycles, and scooters being replaced by their electric equivalents. The technology already exists. Several companies currently make battery-powered scooters for the North American market. At present, electric rickshaws are being produced for many countries around the world. Electric vehicle transportation could already be a reality in many countries. What is missing is the incentive structure to make it happen.

Together, China and India host about one-third of the entire world population. Their economies are some of the fastest growing today. Each has a growing middle class. Under the most likely scenario, a significant increase in the number of cars in these countries is probably inevitable.

Single-passenger electric vehicles would represent a better option for developing countries. Their lower cost, lesser use of metals, easier maintenance, and smaller ongoing energy needs would provide both affordability and environmental benefits. In comparison, fuel cells would require more expensive technology and larger investments in infrastructure.

In March 2009, Tata Motors (India) released a new gasoline sedan, the Nano, which costs only about 100,000 rupees, or less than US$ 2,200. Although it has very good fuel efficiency and low emissions, many believe that it will spell disaster for the planet and the one-billion-people country because it makes gasoline automobiles affordable to so many people. The company is planning to expand to the European and U.S. markets within the next couple of years, offering a version of the Nano below US$ 7,000.

To cater to its growing middle class, China will either import India's cheap automobiles or follow suit with similar models. The country will likely also want to take advantage of international markets. This would also be disastrous for the planet. SPVs might be a better alternative and would have huge potential sales in both the developed and developing world.

The ready market for SPV transportation is actually really, really big internationally. One questions is, do we want a proliferation of gasoline vehicles in the developing world as it seeks better living standards, or do we want conservational and emission-free SPVs? A second one is, who will take advantage of this opportunity first: North America, Europe, Asia? A third one would be, once India and China start delivering cheap gasoline cars to world markets, what are the current leaders in the automobile sector (North America, Europe, Japan, etc.) going to produce and sell, if not the next generation of cars?

Tata Motors is expected to release the Indica EV, a four-passenger electric vehicle, in Europe in 2009 or early 2010! The North

American auto industry might have already lost that battle, being behind developing countries in one of the biggest sector of growth for the future.

15. The Global Environmental Accord: Beyond Kyoto

The Kyoto Protocol is perhaps the world's greatest cooperative effort so far in addressing environmental problems. However, it fails in many ways. As discussed earlier on, it is a regulation-based approach that is somewhat bureaucratic. Secondly, it attempts to address essentially only one issue: greenhouse gases. It does nothing for the conservation of non-renewable metallic resources or many other environmental problems. Thirdly, the intent of the accord is not to stop global warming but only to slow it down. As such, countries are admitting failure from the start.

Why the need for an international accord on global warming? The obvious part of the answer is to force every country to participate and have together a bigger impact on climate change. The not-so-obvious one is that fighting global warming alone would decrease a country's international competitiveness by making domestic production more costly regardless of how it happens, cap-and-trade or a green economic environment. Having a global accord would ensure that countries are treated similarly and continue to compete on an equal footing in both cases.

The Kyoto Protocol

Broadly speaking, the first stage of the Kyoto Accord proposed to have countries reduce their emissions of greenhouse gases for the 2008-2012 period. Groups of countries were given different reduction targets expressed as a percentage of their 1990 emissions levels. For example, the U.S. and Canada had to reduce them to about 6% or 7% lower than they were in 1990.

One of the issues raised with Kyoto is that countries that had underperforming economies in that year have much stiffer standards to meet, their emissions having been lower than usual in 1990. Another problem is national growth levels since then. Countries that had stagnating economies during the 1990s are at a significant advantage compared to others, their emissions having increased less than they would normally have. As well, countries that have seen significant increases in population and economic growth during that period have much further to go to reach Kyoto targets.

Governments that fail to reach the agreed upon targets in 2012 would have to purchase emission credits. That would probably entail the setting up of an international trading system of some kind.

The United Nations Climate Change Conference held in Copenhagen in December 2009 was the latest round of negotiations of the Kyoto Protocol—which is due to expire in 2012. It aimed to produce a successor agreement with much more ambitious greenhouse gas targets. One of its key goals was to have countries agree on a 50% global emission reduction by 2050 (80% for developed countries). The Copenhagen Summit failed in both producing a successor accord and in getting countries to agree on the stiffer targets.

At the last minute, the U.S., China, India, Brazil, and South Africa agreed on a new draft. The proposed agreement has much lower emission limits, is non-binding, and was not adopted by the 193 countries participating in the conference. As such, little has changed with respect to a global accord on greenhouse gases except that the major players at the table have agreed to continue talking.

As discussed earlier, the above is not the best option to address global warming. The whole system is complex, and its structure is essentially regulatory, cumbersome, and bureaucratic for both governments and industry. The GEE could work with the Kyoto Accord and replace cap-and-trade. It would make it easier for industry to reduce greenhouse gases and would remove international competitiveness issues for carbon emissions.

However, a stand-alone international agreement based on a green economic environment strategy and common taxation rates would be more effective in addressing global warming. In addition, it would boost resource conservation, reduce toxic contaminants worldwide, and eliminate international competitiveness issues for all of the above. A GEE-based agreement would be a powerful and comprehensive successor to the Kyoto Accord.

The Global Environmental Accord

The following outlines what an international agreement based on the Green Economic Environment strategy would look like. For the sake of simplicity, it will be referred to as the Global Environmental Accord (GEA) and would generally involve three different groups of countries divided according to GDP levels: high, middle, and low. Of course, this could be expanded as necessary. In some cases, policies would be applicable across the board to all the different groups. In others, different levels would be applied according to purpose and ability.

The Global Environmental Accord would comprise four components: fossil fuels, non-renewable resources, environmental standards, and an export tax to non-GEA members. Initially, the main intent would be to establish a flexible framework that is acceptable to the largest number of countries.

The Fossil Fuel Component

To keep things simple, the following will focus on petroleum, but the GEA would also target other fossil fuels.

Under the Global Environmental Accord, oil would be the object of domestic and import taxes that would raise its cost within individual countries to a certain stable level. The example used earlier was US$ 150/barrel (that might even be too low if the cost of petroleum naturally rises above it). For a market price of $120, the levy would be $30. When the former reaches $140, the tax would drop to $10 so that, over time, oil sells at an average of about $150.

Initial price targets could be chosen so as to approximately achieve the global emission reductions already decided under the Kyoto Protocol, making everything simpler and easier to agree on.

The GEE fossil fuel strategy would call for different price levels for each group of countries in order to maintain the status quo of Kyoto in this respect, i.e. stiffer targets for those better able to afford it.

Higher prices would lead to conservation and promote the growth of the renewable energy sector. For the U.S. and other net importers of fossil fuels, that would mean increased wealth as money would stop hemorrhaging out of their own economies.

After setting the initial target for oil, there would essentially only be a minimal amount of bureaucracy and no emission credit systems to establish. Enforcement throughout the entire business sector would boil down to regular tax collection from a handful of corporations. All a government would have to do is keep track of the international price of oil and establish the appropriate levy to be applied to both local and imported sources at the wholesale level. The tax would be revised periodically as the international cost of oil rises and falls or could be self adjusting.

The business sector would not face a regulation nightmare or need to spend time understanding and implementing arbitrary rules about carbon emissions and a new credit system. No rigid standards would be applied throughout an entire sector. Businesses would face simple input-price decisions for which they already have full expertise. The approach would be dynamic and flexible. Price levels in the accord could be renegotiated every few years in order to meet new emission targets. The tax would not be charged on oil exports so that international prices continue to be set through markets. Decreasing world demand would moderate increases in the price of petroleum.

A full GEE strategy for energy would also tax other fossil fuels, such as coal and natural gas, not only because they are depletable and produce greenhouse gases but also to prevent a shift from one source of emissions to another.

The Non-Renewable Resource Component

The non-renewable resource component of the Global Environmental Accord would be crucial for not only beginning the global conservation of resources but also supporting individual countries' national implementation of their own green economic environment. As already expressed, without multinational agreements a certain amount of conservation could be achieved, but progress would likely reach a ceiling as a result of decreased international competitiveness.

The Global Environmental Accord would establish taxation standards for countries with respect to non-renewable resources, 25%, 35% , 50%, or whatever would be deemed appropriate. Initially, these could be set lower in order to establish a broad-based common framework. They could be raised later to desired levels as economies adjust to the new green economic environment.

The rates would be fixed. Countries would not try to meet a target price as in the case of oil. The levy would be charged on both domestic supplies and imports. This would lead to conservation at home but would not hurt local economies as importing countries would themselves collect it, and consumers would get income and retail tax rebates as per the revenue neutrality principle. Regular international prices would remain in effect for importers, and the system would not penalize resource-poor or developing countries.

The new scheme would, for example, work like this. Richer countries would have to enable full taxation. Poorer ones would have more flexibility, being expected to tax to a minimum of 80% of the established standard. For example, 80% of a 25% tax rate would be 20%. This would promote exports for them and economic growth. The poorest countries would have even greater flexibility. This would let them choose what is best for their particular situation. Of course, nothing would stop them from still electing to meet the full standard as it would maximize conservation.

Higher prices for non-renewable resources would reduce demand and global depletion worldwide while offering poorer and the poorest countries options ranging from implementing the full standard to exercising varying degrees of flexibility in taxation in order to boost their developing economies.

Because GEE taxes would be collected as tariff by importers, the approach would be revenue neutral for individual countries and the money raised would be plowed back into local economies through reduced income and retail taxation. Under the GEA, countries that are resource poor would not be disadvantaged as world prices for raw materials would be maintained, and those having plentiful supplies of minerals would not enrich themselves at the expense of others as has been the case with OPEC for several decades.

The GEA would give us a mechanism to prevent a future of cartelization in which one commodity after another becomes the object of speculation. The OPEC way is not what we want as a future for our children and people around the world.

The Environmental Standards Component

The GEA's environmental standards component would serve to integrate and bring into a common framework existing regulations on pollutants and contaminants. It would mitigate the negative effects of the stricter environmental regulations of some countries on their international competitiveness and promote better national as well as worldwide standards.

Regulatory limits and bans would play an important role in this component. However, the accord would still be primarily market-based, relying on taxation to deter the use of a range of pollutants and contaminants. The component would integrate existing environmental legislation on highly toxic compounds as already agreed upon by the international community. Other chemicals would face worldwide taxation based on their toxicity, bio-accumulation, and potential for cumulative damage to the environment.

Operating on a formula similar to that of the non-renewable resource component—different levels of flexibility applying to groups of countries—the market-based approach would increase environmental standards across the board. Wealthy regions would have to meet full requirements while poorer countries could opt for softer targets as under the other two components. The poorest would have to adhere to less stringent standards although they could still choose to do better and enjoy a cleaner environment.

The Export Tax to Non-GEA Member

Under the GEA, member countries would not be charged export taxes when they import non-renewable raw resources. As explained earlier, a government would impose levies for local production as well as on imports, but exports to international markets would not be taxed for other members of the agreement.

The fourth component of the Global Environmental Accord would give countries the option of charging a levy on raw materials as they are exported to non-GEA members. The tax would range from zero to the full rate charged on imports. No levy would mean no incentive for non-members to join the accord or conserve resources. At the other extreme, the full tax would make non-renewable resources equal in price for both GEA members and non-members, which would result in overall international competitiveness between countries remaining the same. In between, various tax levels would raise prices and result in different degrees of conservation. As such, whether they are participants in the global accord or not, countries would be caught in the Global Environmental Accord's wake, being forced to conserve.

The GEA would not be just another environmental agreement. It would obviously be far more comprehensive. The fourth component of the accord would also give the international community the power to force the conservation of non-renewable resources worldwide.

A View of the Future

In most respects, the Global Environmental Accord would be more flexible and scalable than a regulatory equivalent. It would provide an international framework for the world to not only address global warming but also begin the conservation of non-renewable resources and increase environmental standards across the planet.

The Global Environmental Accord would result in many countries decreasing their dependency on oil. Their renewable energy sector would take off and not look back. Resource conservation would soften the impact of scarcity and delay potential crises. The use of contaminants and pollutants would progressively decrease from the continuous incentive provided by a market-based approach.

The alternative course is the current one: increasingly higher international oil prices, other mineral resources following suit a bit further down the road, environmental contamination getting worse and worse, and global instability growing from poverty and the power shifts associated with scarcity. That future is not that far away. It is probably our children's and grandchildren's!

16. The Environmental Revolution

The GEE provides a mechanism that could trigger an environmental revolution. It could do so because it is revenue neutral and would generally not cost the taxpayer a cent. The massive need for funding required by other environmental strategies is primarily why we have so far failed in making serious environmental progress. In a green economic environment, our habits would have to change, but we would generally enjoy a standard of living similar to the one we are used to.

An environmental revolution in the 21st century would require not only an effective strategy—the GEE—but also a number of key ingredients: fundamental legitimacy, popular support, and timing.

Fundamental Legitimacy

The fight for the environment certainly has undeniable and fundamental legitimacy: stopping global warming, ending the pillage of non-renewable resources, and protecting the environment for ourselves and posterity. The gouging of resources such as metals is a one-way street. There is no second chance!

The Demographics

Politics is a numbers' game. There would be two main demographics involved in an environmental revolution: baby boomers and post-boomers. The former are currently at the helm and will be for some time. They are the ones who have the political power to bring about an environmental revolution. Post-boomers are the new wave of consumers. They are important in keeping companies in business today and will be the ones patronizing them tomorrow.

Together, they have the power and the clout to bring about an environmental revolution.

The Timing

Just a decade ago, people were buying SUVs and gas guzzlers as if there were no tomorrow. Since the escalation of oil prices in 2008, people are slowly waking up to the possibility that the fossil-fuel honeymoon is over. We need to get real about increasing efficiency and shifting to green energy.

Change has started to happen, and global warming is now front and center on the environmental agenda. Unfortunately, contaminants seem to have taken a backseat, and the depletion of most non-renewable resources is not even on the radar screen.

The timing for a revolution is good for two reasons. Firstly, there is a much greater awareness about environmental issues than there was before, hence more political support. Secondly, there is an urgency to act with respect to not only global warming but also contaminants and the depletion of resources.

We now have an environmental strategy that could address all of these issues without costing a cent to taxpayers.

Your Part

Your helping to publicize this book would achieve three things: raise awareness that large-scale change for the environment is possible, let people know that there is a strategy that could make it happen, and support the work we do.

There are a number of ways in which this can be accomplished. Below are several options, but be sure to check the Waves of the Future website (http://wavesofthefuture.net) for updates and other possible strategies. If you have any suggestions, you will also be able to make them through the website.

Libraries

Getting your local and educational libraries to purchase the book is one of the best ways to publicize ideas. It does not cost anything and makes the book available for free to many. Librarians will tell you how to go about it. It is usually just a matter of filling out a form. It can often be done online.

Word of Mouth

Obviously, the easiest way to help with this book is to recommend it to anyone you know. Getting bits and pieces from the media is not going to be enough. People really need to be aware of all of the arguments so that they will not easily be fooled by false claims made by detractors or opponents of environmental change. Political support will be much stronger and more effective if they do.

You can also publicize the book on any available bulletin board. There are some in shopping centers, grocery stores, condominium projects, community centers, at work, etc. Most schools, colleges, and universities have several of them.

You can also involve groups that you belong to, environmental, political, or other. Make sure that you give people the website address so that they can find other ways to participate and the most up-to-date information.

Online Promotion

One of today's most effective ways to advertise and promote something is the Internet. You can use every available opportunity that you have in this respect: blogs, forums, your own website, etc. You can add comments on your MySpace and Facebook pages or on any other network of the kind.

You can also make a YouTube video. You can use search engines to find environmental blogs, forums, discussion groups, etc. and tell people about it. Make sure that you always provide the website address for the book so that people are able to find more information. Do some of these things on a regular basis. Doing it one time only will not be enough.

An obvious means of promotion is to write a review at online bookstore sites such as Amazon.com or where environmental issues are discussed. It does not have to be long but could make a lot of difference. If you are not inclined to writing, just rate it. This is very useful to readers when it comes to purchasing something that they have not already heard about.

Another approach is linking to the Waves of the Future website. Doing this makes it more visible to search engines. In some cases, the website address (http://wavesofthefuture.net/) will suffice and a clickable link will be created automatically. If it is not, the correct HTML coding is:

<a href="http://wavesofthefuture.net/">The 21st Century Environmental Revolution: A Structural Strategy for Global Warming, Resource Conservation, Toxic Contaminants, and the Environment </a>

When you submit your comment, a clickable link should display with the title of the book. If that fails, check the Waves of the Future website or email us. Note that you can change the text of the link (in this case, the title of the book) and should do so if you create many of them. Using several environmental keywords is best.

The News Media

You can bring the book to the attention of the news media. Write to book-review programs on television, columnists, or environmental correspondents. Give newspapers your opinions through their *Letters to the Editor* section. Anyone can do that. Many publications now accept submissions via the web or email.

Best Places to Buy the Book

Royalties to authors can vary hugely depending on where you buy a book. In this case, purchasing directly from the publisher, Waves of the Future (http://wavesofthefuture.net), earns double the regular rate of royalties.

Governments and Corporations

Governments and corporations may be able to help. This is a major project that would benefit society as a whole. Countries may want to participate or support this one way or another.

Supporting the environment has become excellent public relations (PR) for corporations. Here is a chance for them to do something significant. Suggest it to your employer.

Green businesses would hugely profit from the GEE. Supporting this book is probably one of the best investments they could ever make to increase their markets as well as speed up the development of the entire sector. Suggest it to them.

The E-Book Project

The most obvious way to get involved for governments, corporations, and well-off individuals is to participate in the *E-book Project*. An electronic version of *The 21ˢᵗ Century Environmental Revolution* could be made available cheaply or even for free with donations from or the sponsorship of corporations, organizations, and individuals. See the Waves of the Future website for details.

Free digital copies of this book could make a lot of difference in terms of international promotion and political support for the Green Economic Environment and the Global Environmental Accord. The *E-book Project* would benefit everybody. Readers and fans would access the book for free, which would help promote major changes for the environment. Sponsors would get the satisfaction of contributing to a worthy cause. Their donations would go towards meeting their social responsibility goals.

Millionaires and billionaires could sponsor free e-books for everybody without even blinking: from owners of major corporations to investors, celebrities, and musicians. You got connections? Now is the time to use them.

There are thousands of professionals in North America alone, be they medical doctors, professors, engineers, architects, lawyers, computer programmers, etc. It would not take substantial donations from too many of them to reach the goal of free e-books for everybody.

The current population of North America and Europe is over one billion. All that is needed is for one out of every 1,000 people in those wealthy regions of the world to donate to the E-book Project the equivalent of a cup of coffee. Can you spare this much for the environment and your children's future?

Reading Lists

Add the book to online reading lists that you know of. Write to people hosting their own recommended lists and suggest that it be added to them. Some are very popular and would provide a huge amount of publicity for the book.

You

Obviously, *you* are the key to all of this. If *you* assume that *somebody else* will do it, little will ever happen.

Consumer Power

Consumers have an enormous amount of leverage. Everything we buy is a vote we cast. Every time we purchase something, we either reward environmental practices or punish them. We can buy green or purchase wasteful or excessively packaged goods and reward companies that do not care about and destroy the environment. As consumers become more and more aware of their power, attitudes in the business world will have to change. It is up to us. Some of this has already begun to happen, but there is a long way to go.

Our consumption patterns will define our future and that of our children and grandchildren. *Conscious consumerism* can go a long way in making tomorrow's world better. It is one of the most powerful weapons we have today. The GEE would bring in the profound changes that we need for the environment and magnify consumers' power by making green products cheaper and promoting a natural switch to their markets. With it, consumers will be able to choose the kind of world that they want to live in.

Industry as Partner

Industry can be a powerful ally in implementing the GEE. Many sectors would benefit directly from the emerging green markets. Several companies and even major producers of chemicals have already shifted a fair amount of their R&D towards environmental substitutes and rake in a lot of free publicity just for doing the right thing.

Future of Scarcity or Greener Society?

According to most estimates, oil reserves have already peaked or are expected to do so within the next few years. Other mineral resources will follow a similar pattern of increased costs and scarcity as world economies continue to grow. We will start being hit with shortages, prices will rise, and we will face crises as we have with oil. That may be only a few years or decades away.

The U.S. is especially at risk for resource shortages because it is, and has been for a long time, a large manufacturer and producer of goods. To feed its massive industry, it imports minerals from many countries around the world and could face very difficult times ahead. Without the GEE and resource conservation, it will likely have to contend with more crises in the future than it has in the past, resulting in increasing poverty. This is true of many other countries as well.

Both advanced and developing parts of the world have every reason to implement and support a GEE-based global environmental accord to ensure their continued economic success and that of their children. The Global Environmental Accord would prevent the cartelization of one commodity after another. The alternative is the OPEC way and the continuation on a destructive course that would bring only scarcity, crises, and geopolitical turmoil.

The New Landscape

What would the world be like under the GEE? What would the green economic environment imply for us all? Countries would get greener and greener year after year. There would be much less pollution in cities thanks to fuel-cell and battery-powered electric vehicles. SPVs would make most traffic jams a thing of the past, assuming countries opt for non-proliferation policies. Streets would be safer and much less noisy for everybody. Public transportation systems would reduce traffic and be more heavily supported by governments.

Consumers would retain their existing purchasing power, but prices would change to reflect a greener economic incentive structure. Many items would be more expensive, but there would also be more money to spend. Goods would be kept longer and repaired more extensively.

Because of the increase in the price of minerals, consumption would shift towards services and green sectors, promoting growth in various industries. Restaurants, bars, theaters, spas, sports and health centers, personal care companies, etc. would see their business grow. The music, video, and film industries would also do well. Obviously, renewable energy technologies, green construction and home repair, organic farming, etc. would simply thrive under the new system.

Many pollutants (overt and hidden) would be replaced by greener substitutes. Everything we buy would become less harmful to the environment. The waste disposal industry would change dramatically, transforming itself into a source of renewable and non-renewable materials for various industries. Packaging would decrease and become greener and often reusable. We would recycle much more intensively than we currently do.

The green economic environment would reward people for doing the right thing instead of punishing them as is currently the case.

Energy and a Thriving Agricultural Sector

Countries would produce a lot more of their own energy instead of relying on imports. It would come from diversified local sources. This would significantly change balance sheets for oil importing countries, keeping money at home.

Agriculture would become much more lucrative as the demand for its products would increase significantly. As a result, new investment would flow into the industry and serve to boost production, which would in turn help keep prices down.

Governments will likely have to regulate how much good agricultural land can be dedicated to the production of energy. Diversified domestic energy policies would lead to the production of biofuels from non-food crops grown on currently unusable land, refuse, and industrial waste and byproducts. Some energy could come from food crops to eliminate the excess production that often results in depressed agricultural prices and the need for subsidies.

The GEE would lead to a slowdown in the increase of the price of oil, which would have a positive effect on the cost of food. In addition, the system would favor green goods by not taxing them. This would include food.

The New Law of the Land

The Green Economic Environment would transform the world as we know it. The incentives to destroy the environment and waste natural resources would be gone and replaced with ones that would do the exact opposite. Finding ways to improve the environment and stretch out resources—as opposed to destroying them—would become very profitable. The new structure would reshape not only consumer markets but also the industrial sector, which would work for the environment instead of against it.

The companies that produce chemicals would redirect their research and development towards green alternatives to the scores of toxic compounds they themselves invented. The clean up of the environment and of industrial processes would begin and become a thriving sector.

The future would see better living conditions in cities and a robust (and unsubsidized) agricultural sector. The business environment would get greener and cleaner. We would buy fewer material goods and consume more services and environment-friendly products.

Conclusion: Getting It Together

Broadly speaking, there are two main ways to address environmental problems. We can reduce consumption or adopt a *greening strategy*—making what we produce and consume more environment friendly. Both can be equally effective.

The problem with decreasing total consumption is that it calls for a reduction in economic growth, which would mean unemployment and be politically unpopular. Few would support that option at the moment, especially in a context where many countries are still poor.

The GEE would not reduce consumption but make the products we buy greener. It would shift markets away from goods that are unenvironmental. As such, economies could continue to grow and eventually do so in a sustainable way.

Whether they do or not would depend on the strength of implementation of the GEE and on how much greening occurs as a result. Current efforts are not enough. Low GEE levels would not allow for sustainable growth either. Appropriate taxation rates would have to be implemented.

Population growth increases total consumption. If that problem is not addressed, even a powerful strategy such as the GEE may not be able to cope in the future. The equation is simple: more people equals more consumption and pressure on resources. The GEE will only be able to provide for sustainable growth and the improvement of standards of living in the future if the world population is brought under control and the number of people on the planet begins to drop.

There is no question that we need an environmental revolution. Incremental change does not cut it. There is no question that the sooner we start, and the faster we can make it happen, the better. The trillions of dollars' worth of incentive created by the shift to dual-purpose taxation would be a powerful engine capable of bringing about the environmental revolution we need and of placing us into the very midst of the Fourth Wave.

It is really up to us. Baby boomers and post-boomers must join hands in the fight for their own as well as their children's future. This world is ours to make. We have the means, the numbers, and the engine of change.

The New Lifestyles

The GEE would redefine the world we live in. It would change the very structure in which businesses operate. Everybody would have a vested interest in shifting to greener habits, pursuing greener ways, and developing greener products.

The scientific breakthroughs and economic performance of the 20th century have led us to expect unlimited growth and a perpetual bettering of lifestyles. However, part of the prodigious productivity we have experienced so far was derived from fossil fuels, a bountiful and cheap source of energy. Another part resulted from our reckless depletion of resources that are not renewable. Those were and are used as if they are available in infinite supply.

The era of plenty is coming to an end in both cases. The oil crises we have been experiencing are only the first signs of what awaits us down the road. When scarcity hits other minerals, there will be no second chance.

We have made too little headway in solving many of the problems of the last century. We need to do better.

Final Words

With a revenue-neutral Green Economic Environment strategy, we no longer have an excuse not to move ahead with an aggressive conservation and environmental agenda. It is now up to us—baby boomers, post-boomers, voters, and consumers—to choose the world we want to live in, to choose our legacy.

It is time for change.

The next edition of this book will likely not be advertised. Check the Waves of the Future website for availability.

Book II of this series is in the works.
Check the website below for release dates and places to buy.

http://wavesofthefuture.net/

Bibliography

Asbestos. (n.d.). Retrieved on December 13, 2009, from Wikipedia, the free encyclopedia Web site:
http://en.wikipedia.org/wiki/Asbestos

Babcock Gove, P. & the Mirriam-Webster editorial staff (Eds.). (1966). *Webster's third new international dictionary of the English language, unabridged*. Springfield: Merriam.

Bailey, I. (2002). European environmental taxes and charges: Economic theory and policy practice. *Applied Geography, 22(3)*, 235-251.

Baillargeon, J. (2002). *Market and society: An introduction to economics* (M. Galan, Trans.). Halifax, N.S.: Fernwood Publishing.

Beers, B.F. (1986). *World history: Patterns of civilization*. Englewood Cliffs: Prentice-Hall Inc.

The Remarkably Efficient Natural Gas Fuel Cell Car

Blakeslee, T. (2009, September 23). *The remarkably efficient natural gas fuel cell car*. Retrieved November 24, 2009, from EV World Web site:
http://www.evworld.com/article.cfm?storyid=1756

Bruvoll, A. (1998). Taxing virgin materials: an approach to waste problems. *Resources, Conservation and Recycling 22(1-2)*, 15-29.

Common, D. (2008, January 25). *Putting the brakes on ethanol*. Retrieved January 26, 2008, from CBC News: Reports from abroad: David Common Web site:
http://www.cbc.ca/news/reportsfromabroad/common/20080125.html

Contemporary world atlas. (1988). Chicago: Rand McNally and Company.

Council for a Livable World. (n.d.). *U.S. military budget tops rest of world by far*. Retrieved June 05, 2004, from
http://www.clw.org/milspend/ushighestbudget.html.

Diamond, J. (2005). *Collapse: How societies choose to fail or succeed*. New York: Viking, Penguin Group.

Doucet, J.A. (2004). *The Kyoto conundrum: Why abandoning the protocol's targets in favour of a more sustainable plan may be best for Canada and the world*. Ottawa: C.D. Howe Institute.

Dryzek, J. S. (2005). *The politics of the earth: Environmental discourses* (2nd ed.). Oxford, New York: Oxford University Press.

Eberstadt, N. (2004). Population, resources, and the quest to "stabilize human population": Myths and realities. In Huggins, L.E. & Skandera, H. (Eds.), *Population puzzle: Boom or bust?* (pp. 49-65). Stanford, CA: Hoover Institution Press.

Ehrlich, P.R. & Ehrlich, A.H. (2004). *One with Nineveh: Politics, consumption, and the human future.* Washington: Island Press, Shearwater Books.

Environmental Defence. (2007, December 15). Polluted Children, Toxic Nation: A Report on Pollution in Canadian Families. Retrieved September 21, 2009, from:
http://www.toxicnation.ca/toxicnation-studies/reports/toxic-nation-families

Geller, H. (2003). *Energy revolution: Policies for a sustainable future.* Washington, DC: Island Press.

Gerson Lehrman Group (2009, July 16). *Overview of ethanol and flex-fuel auto industry: past & present.* Retrieved October 22, 2009, from:
http://www.glgroup.com/News/Overview-of-ethanol-and-flex-fuel-auto-industry--past--present-41502.html

Houlihan, J., Kropp, T., Wiles, R., Gray, S., & Campbell, C. (2005, July 14). *Body burden—The pollution in newborns: A benchmark investigation of industrial chemicals, pollutants and pesticides in umbilical cord blood.* Retrieved November 11, 2005, from EWG Report Web site:
http://www.ewg.org/reports/bodyburden2/execsumm.php.

Hydrogen fuel cells: Clean but costly. (2004, October). *Consumer Reports Canada, 69(10)*, 18.

International financial statistics yearbook. (2003). [Washington]:International Monetary Fund.

Isidore, C. (2005, November 9). *Big oil CEOs under fire in Congress.* Retrieved November 9, 2005, from CNN/Money Web site:
http://money.cnn.com/2005/11/09/news/economy/oil_hearing/index.htm?cnn=yes

Jackson, T. (2000). The employment and productivity effects of environmental taxation: Additional dividends or added distractions? *Journal of Environmental Planning and Management 43(3)*, 389-406.

Landry, P. (2004, February). *Nicolas Copernicus (1473-1543).* Retrieved November 4, 2005, from Blupete Web site:
http://www.blupete.com/Literature/Biographies/Science/Copernicus.htm

Lean, G. (2004, August 1). *Oceans turn to acid as they absorb global pollution.* Retrieved August 03, 2004, from HealthWorld Online, Healthy News Web site: http://www.healthy.net/scr/news.asp?Id=9602.

Lunder, S. & Sharp, R. (2003, September 23). *Study finds record high levels of toxic fire retardants in breast milk from American mothers: Executive summary.* Retrieved November, 11, 2005, from Environmental Working Group Web site: http://www.ewg.org/reports/mothersmilk/es.php

Lutz, W., Sanderson, W.C. & Scherbov, S. (Eds.). (2004). *The end of world population growth in the 21st century: New challenges for human capital formation and sustainable development.* London, Sterling, VA: Earthscan.

Markowitz, G.E. (2002). *Deceit and denial: The deadly politics of industrial pollution.* Berkeley: University of California Press.

Marx, K. (1957). *Das kapital : Kritik der politischen ökonomie.* Stuttgart: Kröner Verlag.

McIntosh, M. (2000). Globalisation and social responsibility: Issues in corporate citizenship. In Warhurst, A. (Ed.), *Towards a collaborative environment research agenda: Challenges for business and society* (pp. 42-57). New York: St. Martin's Press.

Mercury Policy Project. (2009, September 1). *New studies show mercury in all fish, levels rising in U.S. women.* Retrieved October, 6, 2009, from Mercury Policy Project Web site: http://mercurypolicy.org/?p=778

Nadakavukaren, A. (2000). *Our global environment: A health perspective* (5th ed.). Prospect Heights, IL: Waveland Press, Inc.

Pinderhughes, R. (2004). *Alternative urban futures: Planning for sustainable development in cities throughout the world.* Lanham, MD: Rowman & Littlefield Publishers, Inc.

Pneumaticos, S. (2003). *Renewable energy in Canada. Status report 2002.* Ottawa: Natural Resources Canada.

Polychlorinated dibenzodioxins. (n.d.). Retrieved on December 12, 2009, from Wikipedia, the free encyclopedia Web site: http://en.wikipedia.org/wiki/Polychlorinated_dibenzodioxins.

Revenue Canada. (2002). Final basic table 1—sample data, in *Income Statistics.* Ottawa: Author.

Rocket fuel chemical in California cow milk. (2004, June 22). Retrieved June 22, 2004, from CNN.com, Health Web site: http://us.cnn.com/2004/HEALTH/06/22/milk.chemical.ap/index.html.

Statistics Canada. (2004, March 11). *Employment by industry and sex* [Data file].

Retrieved March 23, 2004, from
http://80-www-statcan-
a.proxy1.lib.umanitoba.ca/english/Pgdb/labor10b.htm.

Stauffer, J. (2008, February 15). *Biocars: Your car may soon be more vegetable than mineral.* Retrieved February 16, 2008, from CBC News In Depth Web site: http://www.cbc.ca/news/background/tech/science/biocars.html

Toffler, A. (1970). *Future shock.* New York : Random House.

Toffler, A. (1980). *The third wave.* New York: W. Morrow.

A unique organisation. (n.d.). Retrieved May 1, 2004, from Scott Bader Commonwealth Web site: http://www.scottbader.com/pub.nsf.

U.S. Department of Commerce. (2002). *Statistical abstract of the United States.* Washington, DC: Bureau of the Census.

Verbeke, A. and Coeck, C. (1997). Environmental taxation: A green stick or a green carrot for corporate social performance? *Managerial and Decision Economics, 18(6),* 507-516.